The Really Wild Guide

The Really Wild Guide to Britain

A Guide to Wildlife Activities for Children

Compiled by
Eric Rowan & David Wallace

*with illustrations by Ann Biggs
and maps by Rodney Paull*

BBC BOOKS

Published by BBC Books,
a division of BBC Enterprises Limited,
Woodlands, 80 Wood Lane,
London W12 0TT

First published 1993
© BBC Enterprises Ltd,
Eric Rowan and David Wallace 1993
ISBN 0 563 36788 1

Maps by Rodney Paull
Illustrations by Ann Biggs

Set in Palatino by BBC Books
Printed and bound in Great Britain
by Butler & Tanner Ltd, Frome
Cover printed by Clays Ltd, St Ives plc

Contents

The Country Code	6
Introduction	7
Key to Maps	9

Wildlife Sites

England	11
Channel Islands	205
Isle of Man	208
Northern Ireland	209
Scotland	217
Wales	241
National Tourist Organisations	265
England's Regional Tourist Boards	266
Wildlife Organisations to Join	268
Gazetteer	273
Index	279

This guide has been compiled from information supplied during 1992 by the wildlife centres themselves. While every effort has been made to ensure that the information is accurate and up-to-date, details are subject to change and we recommend checking information before you set off.

Attractions are signposted by brown tourist signs. Some prices are based on 1992 admission charges and we cannot guarantee that these will not be increased.

The Country Code

Wherever you go it makes sense to act responsibly and follow these practical guidelines which make up the Country Code.

*   Enjoy the countryside and respect its life and work
*   Guard against all risk of fire
*   Fasten all gates
*   Keep your dogs under close control
*   Keep to public paths across farmland
*   Use gates and stiles to cross fences, hedges and walls
*   Leave livestock, crops and machinery alone
*   Take your litter home
*   Help to keep all water clean
*   Protect wildlife, plants and trees
*   Take special care on country roads
*   Make no unnecessary noise

Introduction

If you are an animal-lover, this book is for you. It lists hundreds of places to visit around the country, offering great days out and organised opportunities to see and enjoy the living world around us. From the biggest mammals to the smallest insects, whether they live on land, in the water or in the air, whether wild or domesticated, whether they are natives of these islands or 'settlers', details of where to go, what to see, and how to get there are all to be found in these pages.

The Really Wild Guide to Britain has been compiled by the team behind BBC Television's popular children's natural history series *The Really Wild Show*. It is designed for everyone – young and old alike – who wants to find out more about wildlife.

The book is easy to use. It is divided into geographical sections corresponding to the national and English Regional Tourist Board areas and arranged alphabetically. These sections are divided by counties, with individual sites in alphabetical order.

There are wildlife parks and zoos, nature reserves and National Parks, places of outstanding natural beauty and Sites of Special Scientific Interest. Each entry is packed with essential information about the site and its facilities, including admission charges where applicable, along with a helpful description.

Just as we want to capture the imagination of the millions of viewers who watch our programmes, so we should like to encourage readers to go out and really watch wildlife. We want you to feel involved. So we have made a point of including sites which arrange special events, demonstrations, guided walks and talks, which help visitors understand the animals and their behaviour.

At the end of the book there is a list of organisations to join and put a love of nature into action. And, with an eye to the need for responsible tourism, we have also included the Country Code. The book ends with a gazetteer and an index.

All the information was correct to the best of our knowledge at the time of going to press, but as with all guide books the best advice is to telephone and check the details in advance.

We should like to thank the following members of
The Really Wild Show team for their contributions to this book:
Margaret Black, Mark Brownlow, John Hancock, Kate
Hubert, Ginny Russell, Joanna Sarsby, Martyn Suker and
Michelle Thompson.

Our gratitude also goes to Wendy Hobson, who helped
compile it, and the children's publisher at BBC Books,
Rona Selby.

Finally we should be grateful for your comments on these
sites, and any others you feel we should include in a future
edition of *The Really Wild Guide*. Please send them to this
address:
The Really Wild Guide to Britain
BBC Books, BBC Enterprises
Woodlands, 80 Wood Lane
London W12 0TT

Eric Rowan
David Wallace

The Really Wild Guide to Britain

A welcome publication from BBC Books.
This comprehensive guide from The Really Wild Show *team*
is directed at children but just as useful for anybody
with an interest in wildlife.

The book list Britain's wealth of wildlife parks, zoos, nature
reserves – in fact anywhere that wildlife can be studied and
enjoyed. It is an incentive to all to get out and enjoy our beautiful
British countryside for a week or a day.

William Davis
Chairman English Tourist Board and British Tourist Authority

KEY TO MAPS

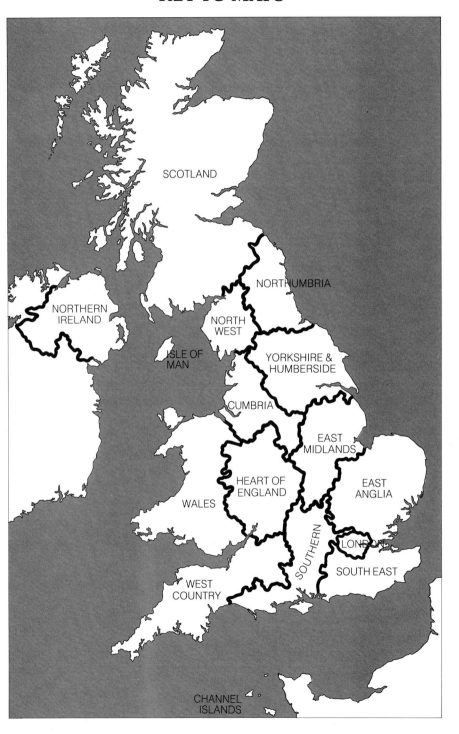

CUMBRIA

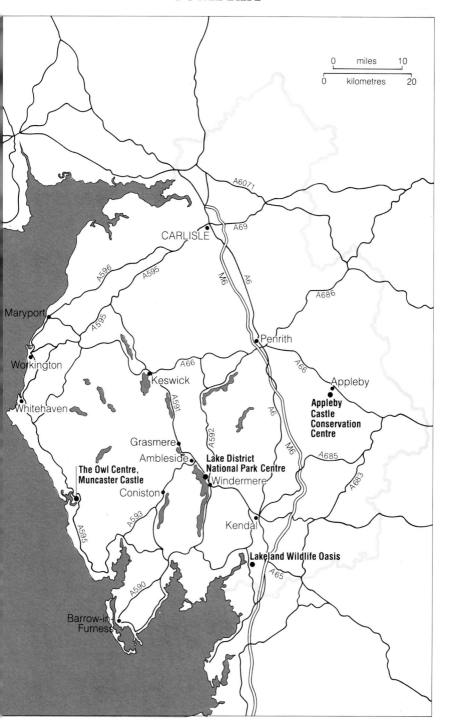

Appleby Castle Conservation Centre

Appleby Castle, Appleby-in-Westmorland, Cumbria CA16 6XH
Tel: (07683) 51402

○ Open daily 10-5 Easter to end Oct.

£ Adults £3.50, children £1.50, OAP's £1.50. Family ticket (2 adults + 2 children) £9.00. Children under 5 go free. Group rates (20 or more) available on request, pre-booking is helpful.

☞ M6 northbound junction 38 or southbound junction 40 (12 miles). A66 Scotch Corner to Penrith road in town centre. Bus: from Penrith and Carlisle. Rail: Scenic Settle to Carlisle line, Appleby station (10 minutes walk).

Facilities: Car and coach parking Toilets Refreshments Gift/book shop Play area.

Facilities for Disabled: Limited wheelchair access.

Restrictions: Dogs on short leads.

Description: The grounds of this beautifully preserved castle provide the tranquil setting for many rare breeds of domestic farm animals – cattle, sheep, goats and pigs – and a large and fascinating collection of waterfowl, owls, pheasants, finches, parakeets and poultry, which can be seen in several ponds and aviaries. There is a nature trail and river walk where free-flying snow geese may be seen. The fine Norman keep and great hall are also open to the public. The centre is approved by the Rare Breeds Survival Trust.

On-Site Activities: Woodland walks in attractively laid-out grounds. Talk on history and Lady Anne Clifford in Great Hall can be pre-booked. Art gallery and studio featuring local watercolour artist within information centre.

Special Events: Jazz festival during the summer; craft displays.

Educational Facilities: School visits.

Lake District National Park Centre

Brockhole, Windermere, Cumbria LA23 1LJ Tel: (05393) 46601

○ Open daily 10-5 Easter to end Oct.

£ Admission free. Car park charges.

☞ On A591 between Windermere and Ambleside.
Bus: regular service from Lancaster and Windermere.
Rail: Windermere station (2 miles).

Facilities: 🅿 Car and coach parking 🚻 Toilets – mother and baby facilities ☕ Refreshments 🛍 Gift/book shop ⛰ Play area.

Facilities for Disabled: ♿ Wheelchair access to shop, café, audio-visual theatres etc.

Restrictions: 🐕 Dogs on leads.

Description: The ideal starting point from which to explore the spectacular Lake District, the largest National Park in England and Wales covering an area of 880 square miles, Brockhole has thirty acres of landscaped gardens, woodland and meadow land in the grounds of an Edwardian house lying on the eastern shore of Lake Windermere. Resident bird life includes woodpeckers, jays, goldcrests and nuthatches, with pied flycatchers, garden warblers and wood warblers in the summer, and pockards and goldeneyes on the lake in winter. There are frequent sightings of red squirrels, roe deer and smaller mammals.

There are nine information centres throughout the Lake District: Bowness Bay, Coniston, Grasmere, Hawkshead, Keswick, Pooley Bridge, Seatolle Barn, Ullswater and Waterhead. Each has local interpretative displays.

On-Site Activities: A full programme of events takes place each year, including many walks and talks with a wildlife theme.

Educational Facilities: Education officer/centre; school visits. A range of tutored programmes are available to schools including educational cruises on Windermere and a series of learning-through-experience modules based at Brockhole and other Lake District locations.

Lakeland Wildlife Oasis

Hale, Milnthorpe, Cumbria LA7 7BW Tel: (05395) 63027

- ○ Open daily from 10 throughout the year except Christmas Day.
- £ Adults £2.85, children £1.25, family £7.50, OAPs £1.75. Group rates (12 or over): adults £2.25, children £1.10, OAPs £1.50.
- ☞ On A6 halfway between Milnthorpe and Carnforth. Near M6 junctions 35 and 36. Bus: to Milnthorpe (2½ miles). Rail: Silverdale station (4 miles).

Facilities: 🅿 Car and coach parking 🚻 Toilets ☕ Refreshments 🛍 Gift/book shop.

Facilities for Disabled: ♿ Wheelchair access.

Restrictions: 🐕 Guide dogs only.

Description: Travel through 3,000 million years of evolution at this unique blend of zoo and museum. By means of working models, hands-on exhibits and living animals themselves, visitors follow the course of life on Earth, starting with the sea, where it all began. Microbes, the simplest of creatures, are viewed through projection microscopes, and sea creatures such as starfish and sea anemones can be seen in the marine aquarium.

The butterfly hall has exotic millipedes, spiders and insects as well as scores of colourful butterflies, and illustrates how life spread on to the land and evolved into a myriad of fascinating forms.

Animals with backbones, such as fish, free-flying birds, fruit bats and other mammals, inhabit the tropical hall where visitors can weigh a whale, design their own mammal or just relax amongst the exotic vegetation.

On-Site Activities: Tours available for groups if pre-booked.

Educational Facilities: School visits.

The Owl Centre

Muncaster Castle, Ravenglass, Cumbria CA18 1RQ Tel: (0229) 717393

○ Open daily 11-5 end Mar to end Oct.

£ Admission to gardens and owl centre only: adults £2.90, children, £1.60, family (2 + 2) £8.00. Group rates: on application. Additional charge for entry to castle.

☞ On A595 west coast road about 1 hour from M6 junction 36. Rail: Ravenglass station (10 minutes walk).

Facilities: 🅿 Car and coach parking ♀♂ Toilets – disabled and mother and baby facilities ☕ Refreshments 👍 Gift/book shop ⛰ Play area.

Facilities for Disabled: ♿ Wheelchair access. Toilets.

Restrictions: 🐕 Dogs on leads.

Description: Muncaster Castle Gardens and Owl Centre boasts one of the finest collections of owls in the world, ranging from tiny scops owls to gigantic eagle owls; it is one of the few places where visitors might see the rare brown fish owl. All the British species are on display in the unique Laybourn aviary. Closed-circuit television has been installed within the nest boxes of some owls, enabling the public to experience the private lives of the owls between February and late April or early May and to watch the incubation of the young. The centre presents a daily meet-the-birds show on summer afternoons when a talk is given on the centre's work. This is a golden opportunity for photographers to take pictures and to ask any questions. Weather permitting, the birds will fly. As well as being the headquarters of the British Owl Breeding and Releasing Scheme to save Britain's endangered barn owls, the centre runs a number of three-day courses designed to give the public a full insight into the world of owls. The centre also organises guided tours, talks, lectures and field work.

On-Site Activities: Guided tours available at extra cost; daily bird of prey flying display; continuous videos; live closed-circuit television in some nest boxes.

Educational Facilities: Education officer/centre; school visits.

EAST ANGLIA

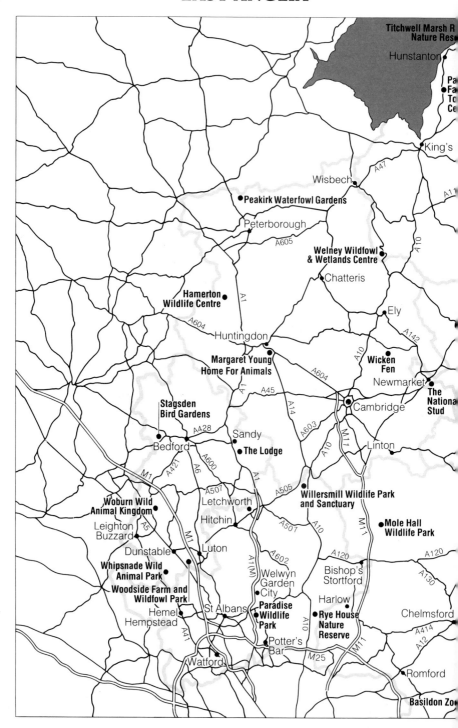

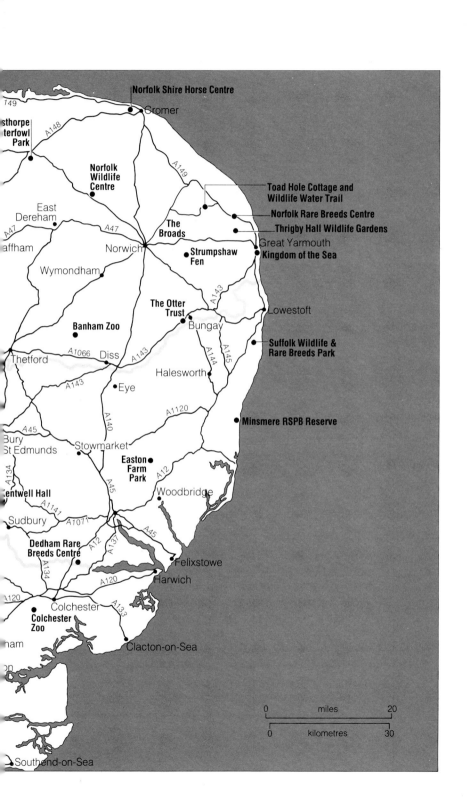

BEDFORDSHIRE
The Lodge

RSPB, The Lodge, Sandy, Bedfordshire SG19 2DL
Tel: (0767) 680551/680541 (shop)

- ○ Open daily 9-9 or dusk. Shop open 9-5 weekdays, 10-5 weekends and bank holidays.
- £ Adults £2.00, children 50p, concession £1.00. RSPB and YOC members free of charge.
- ☞ Off B1042 Sandy to Potton road 2 miles east of Sandy. Bus: United Counties Bedford to Gaminglay bus No. 178. Rail: Sandy station (1 mile).

Facilities: 🅿 Car and coach parking 🚻 Toilets ☕ Refreshments 🛍 Gift/book shop.

Facilities for Disabled: ♿ Wheelchair access.

Restrictions: 🐕 No dogs on nature trails.

Description: There is a birdwatchers' hide, as you would expect at The Lodge, as it is the site of the RSPB's headquarters. The formal gardens and fish ponds give way to mixed woodland with abundant spring flowers. Heath and grassland areas attract tree pipits, the rare natterjack toad and visitors such as nightjars. The hide overlooks a lake which attracts all sorts of birds, dragonflies and squirrels. An abundance of woodland birds can be found here, such as all three woodpeckers and long-tailed tits. The house itself is not open to the public, being full of hard-working RSPB staff!

Educational Facilities: School visits should be arranged with the warden.

Stagsden Bird Gardens

Stagsden, Bedfordshire MK43 8SL Tel: (02302) 2745

○ Open daily 11-6 or dusk throughout the year.

£ Adults £2.50, children (4-15) £1.00, concession £2.00. Group rates: on application.

☞ In village of Stagsden 5 miles west of Bedford on A422. Bus: from Bedford and Milton Keynes.

Facilities: 🅿 Car and coach parking 🚻 Toilets ☕ Refreshments 🛍 Gift/book shop.

Facilities for Disabled: ♿ Wheelchair access.

Restrictions: 🐕 No dogs.

Description: A specialist bird garden and breeding centre set in eight acres of the Bedfordshire countryside, the bird garden has grown from a collection of pheasants (which still figure prominently) and now includes flamingoes, owls and other birds of prey, waterfowl, old breeds of poultry, parrots and mynah birds. Numerous young birds can be seen during the spring and summer months as a result of many successful breeding projects. Hume's bar-tailed pheasant was first bred in captivity here in 1962 and all the world's captive stocks originate from these birds. Stagsden was also one of the first places to breed the cheer pheasant, which has now been reintroduced into Pakistan by the World Pheasant Association. Many of the birds in the collection, such as guans, gallinules and sacred ibis, have their eggs incubated by bantam foster-mothers; birds of prey are also incubated in this way. Although Stagsden has barn owls and a pair of European eagle owls, they do not do free-flying displays.

Several species of crane have recently been added to the collection and specialist breeding programmes are currently under way. To date, the European grey and East African crowned crane have been successfully bred at the centre.

On-Site Activities: Staff are always on hand to answer visitors' questions.

Educational Facilities: School visits.

Whipsnade Wild Animal Park

Dunstable, Bedfordshire LU6 2LF Tel: (0582) 872171

- ○ Open daily 10-6 Mon to Sat, 10-7 Sun and bank holidays in summer, 10-dusk in winter.
- £ Adults £6.95, children £4.95, OAPs £5.60.
  Group rates: adults £5.60, children £3.25, OAPs £4.50.
- ☞ M1 junction 9 or 12 then follow signs.
  Rail/Bus: Luton station; Leisure Link service from Luton or Green Line coach service from London, Victoria.

Facilities: 🅿 Car and coach parking 🚻 Toilets – mother and baby facilities ☕ Refreshments 🛍 Gift/book shop 🛝 Play area.

Facilities for Disabled: ♿ Wheelchair access.

Restrictions: 🐕 No dogs.

Description: Whipsnade Wild Animal Park has approximately 3,000 animals in 600 acres of lush, rolling parkland – about three times more animals than any other collection in Britain. The biggest zoo in Europe, it breeds more large mammals per year than any other collection on the continent – twenty per cent of the world's population of captive-bred cheetahs and fifteen per cent of the world's population of white rhinos. It has provided the major population of Père David's deer reintroduced into China and provided animals for a Przewalskis horse reintroduction programme in Mongolia. Whipsnade keeps the great herd animals extremely well and most of the zoo is divided into huge grazing fields. It is also the base for the great bustard project, a bird which became extinct in Britain two centuries ago. Ten out of the fifteen endangered species of crane breed successfully here, and Whipsnade is the only place in Britain to breed the endangered Indian rhino.

Daily events include the excitement of Whipsnade's birds of prey in flight, when hawks and falcons swoop and dive; sealion splashtime; and meeting the farm animals. The family area and children's farm is one of the best in Britain. The discovery centre has hands-on

displays. miniature deserts and wall-to-wall rainforest –
complete with tarantulas, leaf-cutter ants, giant
constrictors and iguanas. Deer, water deer, wallabies,
marmots and peacocks roam free at Whipsnade. There is
a drive-through area reserved for Asian animals and
viewing mounds and sunken fencing enable visitors to
see wolves almost eye-to-eye.

On-Site Activities: Bird of prey demonstrations; world of sealions; elephants at work; farm talks.

Special Events: Easter egg hunt; steam weekend; craft fair; Fune weekends; teddy bears' picnic; cycling weekend; Christmas at Whipsnade.

Educational Facilities: Education officer/centre; school visits. Tours and talks can easily be arranged for pupils from four years up to degree level on a wide range of topics. For further details please ring (0582) 872171.

Woburn Wild Animal Kingdom

Woburn Park, Woburn, Bedfordshire MK17 9QN Tel: (0525) 290407

○   Open daily from 10 early March to end Oct.

£   Adults £7.00, children and OAPs £4.50.
    Group rates: on application.

☞   From M1 junction 12 or 13 or from A5 turn at Hockliffe on to A5130 (A50).

Facilities: 🅿 Car and coach parking 🚻 Toilets – disabled and mother and baby facilities 🍽 Refreshments 🛍 Gift/book shop ⛰ Play area.

Facilities for Disabled: ♿ Wheelchair access Toilets.

Restrictions: 🚫 No dogs. No soft-top cars and no barbecues.

Description: For more than a century, Woburn has been a sanctuary for wildlife. Over that period, the Dukes of Bedford have grazed a whole variety of wild animals including zebra, giraffe, wildebeest, moose, pampas deer, musk ox, bison and many others. But perhaps Woburn is best known for saving from extinction Père David's deer which lived in China. From only a handful of animals surviving in collections around the world, Woburn built up a herd of several hundred, and these deer are now found in numbers around the world. Today the great deer park and the stately home of Woburn Abbey are separate from the Wild Animal Kingdom, a traditional sarari park which includes a lion enclosure; hippos wallowing in a muddy lake; white rhinos; a huge field containing zebra, eland, crowned crane and giraffe; bongos, monkey jungle; bears and Asian elephants. Unlike many other such parks, tigers are not kept caged. A cable-car ride ends in pets corner where there are llamas, donkeys, parrots and farmyard animals to be fed and stroked.

On-Site Activities: Regular animal display times, pets corner, adventure playground.

Educational Facilities: School visits.

Woodside Farm and Wildfowl Park

Mancroft Road, Aley Green, near Slip End, Luton, Bedfordshire LU1 4DG
Tel: (0582) 841044

○ Open Mon to Sat 8-5.30 throughout the year. Closed Sun and Christmas holidays.

£ Adults £1.40, children £1.10. Group rates: £1.00

☞ M1 junction 9 then follow signs.
Bus: Hop'n'Stopper from Luton bus station.

Facilities: 🅿 Car and coach parking 🚻 Toilets – mother and baby facilities ☕ Refreshments 🛍 Gift/book shop 🛝 Play area.

Facilities for Disabled: ♿ Wheelchair access.

Restrictions: 🐕 No dogs.

Description: Woodside Farm boasts a large range of farm animals and wildfowl including pure and rare breed chickens, ducks and waterfowl, guinea fowl, pheasants, turkeys, geese, swans, owls, peacocks, pygmy goats, rabbits and guinea pigs. Among the larger animals on view are horses, donkeys, deer, sheep, cattle and pigs. There are opportunities for children to get up close to some of their favourite animals and ride on tractors and steam engines. There are swings, a Tarzan trail, swing bridges and a fort. Fresh produce, dairy products and pet foods are available from the traditional farm shop. A craft and gift centre and a blacksmith's shop are also on site.

On-Site Activities: Guided tours.

Special Events: Events organised and advertised throughout the year.

Educational Facilities: Education officer/centre; school visits.

CAMBRIDGESHIRE
Hamerton Wildlife Centre

Hamerton, Huntingdon, Cambridgeshire PE17 5RE Tel: (08323) 362

- ○ Open daily 10-6 in summer, 10-4 in winter. Closed Christmas Day.
- £ Adults £3.50, children, (5-14 years old) £2.00, (children under 5 free), OAP's £3.00. Group rates: 20% discount for parties of 15 or more, but these must be pre-booked.
- ☞ Follow signs off A1 at Sawtry and the A14 at the B660 interchange.

Facilities: 🅿 Car and coach parking 🚻 Toilets ☕ Refreshments 🛍 Gift/book shop 🎪 Play area.

Facilities for Disabled: ♿ Wheelchair access.

Restrictions: 🐕 No dogs. Children must be accompanied by an adult.

Description: The centre is set in sixteen acres of parkland in the Huntingdonshire wolds. Many of the animals here are rare or endangered and part of co-ordinated captive breeding programmes. Primates include Madagascan lemurs, tiny marmosets and acrobatic gibbons. The only group of South American sloths to breed in the UK is found here. The collection includes: otters, wallabies, Scottish wildcats, meerkats, deer, and over 100 species of bird, many of which you cannot see elsewhere in the UK. Guided tours are provided if arranged in advance, and there are regular falconry displays. There is a covered picnic area.

On-Site Activities: Guided tours available if pre-booked.

Special Events: Falconry demonstrations; special bank holiday weekends; car rallies. Full information from the centre.

Educational Facilities: School visits.

Margaret Young Home for Animals

Wood Green Animal Shelters, London Road, Godmanchester, Cambridgeshire PE17 5LJ Tel: (0480) 830014

- ○ Open daily 9-3 throughout the year.
- £ Admission free.
- ☞ Off A604 (adjoins M11/A11) to Godmanchester; signposted on A1198. Rail: Huntingdon station.

Facilities: 🅿 Car and coach parking 🚻 Toilets – mother and baby facilities ☕ Refreshments 🛍 Gift/book shop.

Facilities for Disabled: ♿ Wheelchair access.

Description: The Wood Green Animal Shelters comprise a Victorian house in Wood Green, a derelict pig farm in Heydon, Herefordshire and the Margaret Young Home for Animals at Godmanchester. Being one of Europe's most progressive animal charities, there are many different species of animals to see, both wild and domesticated. They include Hissing Sid, the one-winged swan, and Luchi the llama. Both these animals live at the fifty-acre Godmanchester site. This houses the most varied range of species from hamsters to horses. The other two sites deal mainly with domestic animals. All the centres have one thing in common: they offer treatment to animals whose owners suffer financial hardship.

On-Site Activities: Guided tours available on request; pet care centre; picnic areas.

Special Events: Family fun days in August; tortoise day in September; car boot sale in August.

Educational Facilities: Education officer/centre; school visits.

Peakirk Waterfowl Gardens

Peakirk, Peterborough, Cambridgeshire PE6 7NP Tel: (0733) 252271

- ○ Open daily 9.30-6.30 Apr to Sep, 9.30-dusk Oct to Mar.
- £ Adults £2.50, children £1.25, family £5.00, concession £2.00. Group rates: adults £2.00.
- ☞ On A15 Glinton road north of Peterborough then east for 1 mile on B1443.

Facilities: 🅿 Car and coach parking 🚻 Toilets 🍽 Refreshments 🛍 Gift/book shop.

Facilities for Disabled: ♿ Wheelchair access; free wheelchair loan.

Restrictions: 🐕 No dogs.

Description: Seventeen acres of ponds, open grass, woodland and gardens interconnected by paths provide a home for 157 species of waterfowl. Native birds live alongside more exotic species such as the flamingo and crane; there are around 700 birds to be spotted. Several species of swan can be found including the black-necked, the Bewick and the boisterous trumpeter. Hawaiian, barnacle and pink-footed geese are periodically visited by passing migrants. A number of ducks such as the whistling, eider and shell can also be identified. A newly constructed hide provides a comfortable shelter in which to view the numerous wildfowl. However, most of the birds are tame and can be approached in safety.

On-Site Activities: Guides, walks and talks by prior arrangement.

Special Events: Special activity days during half term and summer holidays.

Educational Facilities: Education centre; school visits.

The Welney Wildfowl & Wetlands Centre

The Wildfowl & Wetlands Trust, Welney Centre, Hundred Foot Bank, Welney, near Wisbech, Cambridgeshire PE14 9TN Tel: (0353) 860711

○ Open daily 10-5 throughout the year except Christmas Eve and Christmas Day.

£ Adults £2.95, children £1.50, family £7.40, OAPs £2.20.

☞ Follow signs from A10. Located 12 miles north of Ely, 26 miles north of Cambridge, 33 miles east of Peterborough. Rail: Littleport station (8 miles).

Facilities: P Car and coach parking Toilets Refreshments Gift/book shop.

Facilities for Disabled: ♿ Wheelchair access limited to main observatory.

Description: Welney is a wild nature reserve and one of Europe's most important conservation areas. Visitors can explore this vast unspoilt remnant of original fenland and get close to nature thanks to a network of twenty-one camouflaged observation hides.

This is the winter habitat of tens of thousands of migrant birds, especially famed for Bewicks and whoopers which perform a 'swan spectacular' floodlit on the main lagoon, with hundreds of birds feeding and fighting, courting and preening, in a breathtaking display.

In the summer, walks can be taken across the reserve in the surroundings of wetland wild flowers and taking in the reedbed boardwalk and pond-dipping area.

On-Site Activities: Guided tours and afternoon classes available. Full range of events throughout the year; contact the centre for further details.

Special Events: Dawn chorus walks; floodlit swan evenings (may need to book).

Educational Facilities: Education officer/centre; school visits. Teacher packs, activity books, pondwatch and follow-up projects available.

Wicken Fen

Lode Lane, Wicken, Ely, Cambridgeshire CB7 5XP Tel: (0353) 720274

- ○ Open daily dawn to dusk throughout the year except Christmas Day.
- £ Adults £2.50, children £1.25. Group rates: adults £1.90, children 90p. NT members free of charge.
- ☞ A1123 linking A10 Cambridge to Ely and A142 Newmarket to Ely. Following signs, turn right at western end of village. Bus: Sun service between Ely and Cambridge May to Sep. Very limited service during the week.

Facilities: P Car and coach parking ♦♦ Toilets - mother and baby facilities ☕ Limited refreshments.

Facilities for Disabled: ♿ Wheelchair access; three-quarters continuous boardwalk trail ideal for wheelchairs.

Restrictions: 🐕 Dogs under strict control.

Description: This survives as an undrained remnant of the great East Anglian levels. It is incredibly rich in plant and animal life, both on land and in the water. There is a whole range of habitats, from open water through to reedbeds, sedge fields, to scrub woodland. You can follow marked trails around the fen, and observe birds from the three hides. Much of the trail is boardwalks, giving easy access on what could otherwise be very boggy ground. The eagle-eyed may spot many of the twenty-three species of butterfly, and eighteen dragon and damselflies to be found here. There is a great number of rare birds; use the checklist available and see how many you can spot. If you're lucky you might see hen and marsh harriers, barn owls, hobbys, bearded reedlings, snipe and woodcocks. Wicken Fen is a well-managed site, preserving habitats and species almost lost to East Anglia.

On-Site Activities: Occasional guided walks for individuals. Parties can book evening walks May to September.

Educational Facilities: Education officer/centre; school visits. Resource information available from Fen Cottage open Sundays and bank holiday Mondays 2-5 April to October.

East Anglia – Essex 29

ESSEX
Basildon Zoo

Vange, Basildon, Essex SS16 4QA Tel: (0268) 553985

○ Open weekdays 10-4, weekends 10-5 or dusk throughout the year except Christmas Day and Boxing Day.

£ Adults £2.50, children £1.25. Group rates: 10% discount.

☞ M25 junction 29 A127 or junction 30 A13, follow signs to zoo 10 miles from Dartford Tunnel A13. Bus: to Basildon or Pitsea stations. Rail: Basildon or Pitsea stations (2 miles from each).

Facilities: 🅿 Car and coach parking 🚻 Toilets – disabled facilities ☕ Refreshments 🛍 Gift/book shop 🛝 Play area.

Facilities for Disabled: ♿ Wheelchair access and paths wide enough for wheelchairs. Toilets.

Restrictions: 🐕 No animals.

Description: This is a small zoo set in pretty gardens of trees and shrubs, but does boast a café, adventure playground and play area with large sand-pit as well as two picnic areas. There is a varied collection of birds including macaws, owls, sarus cranes, buzzards and night herons. Big cats such as lions and pumas are to be found, as are monkeys such as the tiny marmosets and squirrel monkeys. You can also meet llamas, porcupines, pot-bellied pigs, racoons and wallabies. Visitors are allowed to feed some of the animals with specially bought food and to touch them in the pet-patting area.

On-Site Activities: Talks and demonstrations with the animals; questions and answers with the public.

Educational Facilities: Education centre under development; school visits.

Colchester Zoo

Maldon Road, Stanway, Colchester, Essex CO3 5SL Tel: (0206) 330253

- ○ Open daily from 9.30 throughout the year except Christmas Day.
- £ Adults £5.00, children £3.00, OAPs £4.00.
  Group rates: on application.
- ☞ From A12, take A604 exit south of Colchester and follow signs. Bus: Colchester bus station.
  Rail: Colchester station (4 miles).

Facilities: 🅿 Car and coach parking 🚻 Toilets – mother and baby facilities ☕ Refreshments 🛍 Gift/book shop 🛝 Play area.

Facilities for Disabled: ♿ Wheelchair access although there are some steep hills which can be difficult; special route available; wheelchair hire can be booked.

Restrictions: 🦮 Guide dogs only.

Description: Colchester Zoo has a large collection of rare and exotic animals, about 150 species, with most of the popular species represented: elephants, lions, Siberian tigers, bears, chimpanzees, penguins as well as tiny marmosets and tamarin monkeys. It has some of the best collections of cats and primates in the country.

Fifteen species of cat include the rarely seen leopard cat of eastern Asia as well as ocelots, jaguars, snow leopards, panthers and the Siberian tigers in Tiger Valley. Chimpanzees can be seen in a large new chimpanzee house which has plenty of climbing ropes and trees and a spacious outdoor area complete with artificial termite mound. Another new development is Out of Africa, a walk-through area featuring mangabeys and colobus monkeys as well as endangered lion-tailed macaques. There are white rhinos in a large outside pen and a wolf wood for Canadian timber wolves. There is a children's farm area whose animals are labelled 'familiar friends', although more unfamiliar ones are zeedonks, the strange offspring of a zebra and a black Arabian ass, claimed to be unique in the world.

On-Site Activities: Daily programme of displays include feeding the elephants, snake-handling, falconry, penguin parade, parrot, seal and sealion demonstrations.

Educational Facilities: School visits. Educational information available. Talks, worksheets and project information available for schools in addition to usual daily events.

Dedham Rare Breeds Centre

Mill Street, Dedham, Colchester, Essex CO7 6DR Tel: (0206) 322176

○ Open daily 10-4.30 Mar, 10-5.30 Apr to end Oct, 10-3.30 Nov to mid-Jan, closed mid-Jan to end Feb.

£ Adults £3.00, children £1.80.
   Group rates: on application.

☞ Take A12 south from Ipswich to Dedham village then follow signs for public car park; the centre is next to the car park.

Facilities: 🅿 Car and coach parking 🚻 Toilets – disabled and mother and baby facilities ☕ Refreshments 🛍 Gift/book shop 🎠 Play area.

Facilities for Disabled: ♿ Wheelchair access.

Restrictions: 🐕 No dogs.

Description: This comprehensive collection of rare farm animals is a must for all rare breeds enthusiasts. All are British farm livestock which were once common on farms but are now in danger of extinction. Today, many of these breeds would be extinct if it were not for the work carried out by the centre and the Rare Breeds Survival Trust. This provides an excellent environment for young and old alike to study many subjects, especially conservation and the environment. The farm is set in sixteen acres of the Dedham dales and the enclosures are divided into large paddocks. This is an area of outstanding natural beauty, known from the paintings of landscape artist John Constable. The farm's large collection of animals includes the Gloucester Old Spot pig and many cattle and sheep, most of which can be touched and fed. Chickens and Norfolk turkeys scratch for food around the buildings.

On-Site Activities: Free guided tour for all booked groups.

Special Events: Annual sheep shearing, dipping and lambing.

Educational Facilities: Education officer/centre; school visits.

Mole Hall Wildlife Park

Widdington, near Saffron Walden, Essex CB11 3SS Tel: (0799) 40400

○ Open daily 10.30-6 or dusk throughout the year except Christmas Day.

£ Adults £3.50, children £2.25, concession £2.00. Group rates: £2.00.

☞ From the south, go to M11 junction 8 then take A120 for 1/2 miles towards Puckeridge. Turn right on to B1383 and follow signs 5 miles on the right. From the north, go to Stump Cross and take B1383 towards Newport then follow signs 1/2 miles south of Newport.

Facilities: 🅿 Car and coach parking 🚻 Toilets ☕ Refreshments 🛍 Gift/book shop 🎪 Play area.

Facilities for Disabled: ♿ Wheelchair access.

Restrictions: 🐕 Dogs only in car park.

Description: Mole Hall is set in the twenty-acre grounds of a thirteenth-century moated manor house. It is home to two species of otter and claims to be the first regular breeder in this country of the North American otter. Its wide variety of animals range from South American Lama Guanaco to Kenyan eagle owls, serval cats and Formosan sika deer, which are extinct in the wild. Other animals include chimpanzees, coatis, Vietnamese pot-bellied pigs and a number of domesticated breeds. Visitors can also walk through the jungle-like butterfly pavilion.

On-Site Activities: Field walk; talks by arrangement; on-site entomologist and guide; otter feeding daily.

Educational Facilities: Education officer/centre; school visits. Teachers' packs and worksheets available.

HERTFORDSHIRE
Paradise Wildlife Park

White Stubbs Lane, Broxbourne, Hertfordshire EN10 7QA
Tel: (0992) 468001

- ○ Open daily 10-6 or dusk throughout the year.
- £ Adults £3.50, children £2.50, OAPs £3.00. Group rates (over 10): 10% discount.
- ☞ Off A10 on A1170.

Facilities: 🅿 Car and coach parking 🚻 Toilets – disabled and mother and baby facilities ☕ Refreshments 🛍 Gift/book shop ⚠ Play area.

Facilities for Disabled: ♿ Wheelchair access; ramps. Toilets.

Restrictions: 🐕 Dogs on leads in play area only, not in zoo.

Description: The leisure and wildlife park is set in picturesque Broxbourne Woods. This seventeen-acre family-run complex has a wide variety of farmyard and exotic animals ranging from pigs, calves and lambs to lions, monkeys, camels and zebras. Daily activities include walking the camels, guided tours and pony rides. Paddock animals are fed in the morning and afternoon. The lions are fed six days out of seven at five. Other facilities include a woodland railway, a woodland walk, a walk-through aviary, children's rides, an adventure playground, crazy golf and an education centre. Animal feeding is encouraged and visitors can purchase prepared feed pots to feed the paddock and domestic animals.

On-Site Activities: Guided tours; walking the camels; pony rides; animal feeding.

Educational Facilities: Education officer/centre; school visits.

Rye House Nature Reserve

Rye House, Rye Road, Hoddesdon, Hertfordshire EN11 0EJ
Tel: (0992) 460031

○ Open daily 9-5 throughout the year except Christmas Day.

£ Adults £1.50, children 50p.
RSPB and YOC members free of charge.

☞ From Hoddesdon (off A10) follow signs to Rye House. Rail: Rye House station (300 metres).

Facilities: 🅿 Car and coach parking 🚻 Toilets 🍽 Refreshments.

Facilities for Disabled: ♿ Wheelchair access.

Restrictions: 🐕 No dogs. No bicycles.

Description: This is the closest RSPB reserve to London. It boasts a spectacular range of birds. A number of paths and hides allow visitors to get the best views of the birds in a variety of natural and man-made habitats. The reedbeds are visited by cuckoos and reed warblers in the summer, whilst the very rare bittern often rests here in the winter. The willow scrub habitat attracts woodpeckers and six species of warbler in the autumn, while the goldfinch and bullfinch are often seen here in the summer. The lakes teem with wildfowl throughout the year, but particularly in the winter months when gadwall, shoveller and tufted ducks are a common sight. Visitors can get excellent views of kingfishers nesting from one of the many hides in the reserve. A large inland colony of common terns breed very successfully on specially designed rafts which float on the lakes. Green sandpipers fly in during the autumn before continuing their migration south.

On-Site Activities: Guided walks and talks, courses for school groups available. Events calendar available on request.

Special Events: Information on request.

Educational Facilities: Education officer/centre; school visits. Contact the reserve to book a school visit.

Willersmill Wildlife Park and Sanctuary

Station Road, Shepreth, near Royston, Hertfordshire SG3 6PZ
Tel: (0763) 262226

○   Open daily 10.30-6 Mar to Oct, 10.30-5 in winter.

£   Mon to Sat: adults £3.25, children £1.75, OAPs £2.50. Sun and bank holidays: adults £3.75, children £2.00, OAPs £3.00. Group rates (over 14): 20% discount if paid 3 weeks in advance, 10% discount if paid on the day.

☞   Off A10 London road between Cambridge and Royston. Bus: from Cambridge and Royston. Rail: Shepreth station.

Facilities: P Car and coach parking ⴄ Toilets 🍽 Refreshments 👍 Gift/book shop ⛰ Play area.

Facilities for Disabled: ♿ Wheelchair access.

Restrictions: 🐕 No dogs.

Description: The wildlife sanctuary is set in the beautiful countryside of Hertfordshire. A fish farm is home to large carp, some of which are so tame they will feed off your finger. The sanctuary's collection of animals and birds is very friendly: many will approach visitors and allow themselves to be stroked. Some animals are allowed to roam free, such as squirrels, foxes and rabbits. There is also a variety of water birds, and a wide range of mammals including wolves, monkeys, coatis, pine martens, racoons, deer and porcupines. The sanctuary houses a tropical house, and of special interest to children is a farm exhibit where they can encounter farmyard animals face to face.

On-Site Activities: Guided walks at additional cost.

Educational Facilities: School visits.

NORFOLK
Banham Zoo

The Grove, Banham, Norfolk NR16 2HB Tel: (0953) 87771

- ○ Open daily 10-6.30 or dusk throughout the year except Christmas Day and Boxing Day.
- £ Adults £5.00, children £3.00, concession £2.50. Group rates: adults £3.50, children £2.50.
- ☞ On B1113 Norwich to Bury St Edmunds road between Attleburgh and Diss. Bus: Eastern Counties bus from Norwich to Kenninghall.

Facilities: 🅿 Car and coach parking 🚻 Toilets – mother and baby facilities ☕ Refreshments 🛍 Gift/ book shop 🎪 Play area.

Facilities for Disabled: ♿ Wheelchair access.

Restrictions: 🐕 No dogs.

Description: Banham Zoo is set in over fifty acres of countryside and is home to a large collection of owls and other animals including monkeys and snow leopards. The woodland walk will take you past the flamingoes and monkey jungle island. Many species here are endangered in the wild. Visit the reptile house and Australian paddock and also find the parrots, meerkats, penguins, fur seals, zebras, maned wolves, chimpanzees, camels, llamas, otters and maras! Watch out for the animals' feeding times. Younger visitors can get close to more familiar animals at the pets and farmyard corner. Refreshments are available at the bistro, and you can sample Banham's own home-made cider from their award-winning press. In bad weather you can always enjoy the indoor activity centre, and outside children will have fun on the adventure playground.

On-Site Activities: Keeper walks and talks: Perky Parrot's story room relating the story of rainforests.

Special Events: Special events throughout the year; themed animal weeks and special days for children.

Educational Facilities: Education officer/centre; school visits.

The Broads

*Broads Authority, Thomas Harvey House, 18 Colegate, Norwich
Norfolk NR3 1BQ Tel: (0603) 610734*

Description: The Broads is a wetland area with over 124 miles of navigable rivers and lakes and is Britain's most extensive inland waterway system. There are over forty Broads, fourteen of which provide exceptional sailing opportunities.

The Broads is renowned for its wildlife and encompasses three National Nature Reserves (Ludham Marshes, Bure Marshes and Hickling Broad). In the upper valleys there are wide areas of fens, home to swallowtail butterflies, cormorants, bitterns and marsh harriers. Downstream, open expanses of grazing land provide a Dutch-style scenery of scattered windpumps (many of which are open to the public). Numerous bird species live in the area especially in winter. The floating visitor centre at Ranworth Conservation Centre near Wroxham has a permanent exhibition detailing development and problems facing the Broads. It also provides excellent opportunities for birdwatching.

Kingdom of the Sea, Great Yarmouth

Marine Parade, Great Yarmouth, Norfolk NR30 3AH Tel: (0493) 330631

○   Open daily 10-dusk throughout the year except Christmas Day.

£   Adults £4.25, children and OAPs £3.25, family (2 + 2/3) £12.50. Group rates: on application.

☞   A12 or A47 to Great Yarmouth. Bus: to Great Yarmouth. Rail: Great Yarmouth station (2 miles).

Facilities: 🚻 Toilets – disabled and mother and baby facilities ☕ Refreshments 🛍 Gift/book shop.

Facilities for Disabled: ♿ Wheelchair access. Toilets.

Restrictions: 🐕 Guide dogs only.

Description: One of Britain's largest all-year, all-weather marine aquariums, Kingdom of the Sea displays British marine life from along the north coast of England in a natural environment. The visitor can see the graceful rays and flat fish playing hide and seek in the island sands exhibit and waves lapping over a seashore setting named Winterton Dunes. In the ocean tunnel, tropical reef sharks such as nurse, tiger and black-tipped, swim silently overhead and hundreds of tropical fish can be viewed through the portholes of the Green Submarine or the coral canyon. Talks are given throughout the day.

On-Site Activities: Talks throughout the day. Film showing in audio-visual room. Introductory talks for group bookings. Pirate trails, colouring sheets and fun quizzes on sale in reception area.

Special Events: Children's Saturday club run through the autumn and winter.

Educational Facilities: Education officer; school visits. Teachers' resource packs are available on request.

Norfolk Rare Breeds Centre

Decoy Road, Ormsby St Michael, Great Yarmouth, Norfolk NR29 3LY
Tel: (0493) 732990

- ○ Open Sun to Fri 11-5 Easter to end Oct, Sun only in winter.
- £ Admission free.
- ☞ Just off A149 North Walsham to Great Yarmouth road. Bus: Telephone for information as buses are seasonal.

Facilities: 🅿 Car and coach parking 🚻 Toilets ☕ Refreshments 🛍 Gift/book shop.

Facilities for Disabled: ♿ Wheelchair access.

Restrictions: 🐕 No dogs.

Description: A small and interesting collection of rare breeds is kept at the centre. There are fourteen varieties of sheep, nine of pigs, six of cattle and various goats. The saddleback, Tamworth, Gloucester Old Spot, British lops and Oxford sandy and black pigs are on the category 1 endangered list. Surrounded by the Norfolk Broads in a quiet and rural setting, otters can occasionally be seen in the main dyke as well as kingfishers, herons, partridges and pheasants. Small muntjac deer have been sighted in the evening. The centre, which covers a ten-acre site, is not commercialised in any way. The centre is happy to arrange guided tours, walks, horse and cart rides and nature walks, which are usually included in group visits.

On-Site Activities: Guided tours and walks. Horse and cart rides and nature walks are usually included in group visits.

Educational Facilities: School visits. Worksheets for children.

Norfolk Shire Horse Centre

West Runton Stables, West Runton, Cromer, Norfolk NR27 9QH
Tel: (0263) 837339

○ Open Sun to Fri 10-5 Easter to end Oct.

£ Adults £3.00, children £2.50, OAPs £2.00.
Group rates: on application.

☞ On A149 between Cromer and Sheringham. Bus: from Norwich to West Runton via Cromer. Rail: Eastern Region Norwich to Sheringham or West Runton stations (2 miles/½ mile).

Facilities: ▣ Car and coach parking ♔ Toilets
▣ Refreshments ♣ Gift/book shop ▲ Play area.

Facilities for Disabled: ♿ Wheelchair access.

Restrictions: 🐕 Dogs on leads.

Description: Many types of heavy horse can be seen here as well as the Shire. The public is invited to meet and learn about such ancient breeds as the Suffolk Punch, Clydesdale and Percheron. The horses can be seen at work in twice-daily displays of ploughing and log-hauling, muck-spreading and potato-lifting. The farrier visits once a week to reshoe the horses and ponies, and you can watch him at work. There are other types of smaller native ponies such as the Connemara, Highland, Dales, Fell, Welsh, Dartmoor, Exmoor and New Forest together with farm animals and poultry to give a feel of farms in the past when horse-power ruled. A collection of old farm equipment and rural objects adds to the sense of history.

    The horses and ponies all breed here and in the summer visitors get a chance to meet the foals. Youngsters will also like meeting the turkeys, pheasants, ducks, goats and pot-bellied pigs. They can play in the adventure playground and refreshments are available in the café and shop.

On-Site Activities: Talks and demonstrations twice daily.

Special Events: Frequently arranged throughout the season.

Educational Facilities: School visits.

Norfolk Wildlife Centre
Great Witchingham, Norwich, Norfolk NR9 5QS Tel: (0603) 872274

○ Open daily 10.30-6 or dusk Apr to end Oct.

£ Adults £3.50, children £2.00. Children free during Jul and Aug and on Sat throughout the year (this does not apply to party rates). Wheelchair-bound visitors and those pushing wheelchairs free.
Group rates: adults £2.50, children £1.50.

☞ Well signposted, 11 miles north-west of Norwich on A1066 Norwich to Fakenham Road.
Bus: Eastern Counties Nos. 450, 454, 456.

Facilities: 🅿 Car and coach parking 🚻 Toilets ☕ Refreshments 🛍 Gift/book shop 🅰 Play areas.

Facilities for Disabled: ♿ Wheelchair access. Free admission for wheelchair-bound visitors and those pushing the wheelchair.

Restrictions: 🐕 Guide dogs only.

Description: Forty acres of undulating parkland provide the setting for an abundance of rare and unusual trees and flowering shrubs. The animals are kept in large, open enclosures under semi-natural conditions. Most of the animals are native to Europe, although some species from other parts of the world are also exhibited, including wallabies, Barbary apes, coypus and Arctic foxes. A large wild heronry is located in the centre of the park, which is home to over thirty pairs of nesting birds. In early spring, visitors can watch the herons coming in to their nests to feed their chicks.

There is a model farm with tame farm animals, a pets corner and two play areas for children. A team of trained reindeer pull a light car round the park on summer afternoons. The Norfolk Wildlife Centre specialises in breeding endangered European birds and mammals and has successfully reintroduced eagle owls, cheer pheasants, barn owls, little owls, badgers and otters into the wild.

Educational Facilities: Education officer/centre; school visits.

Park Farm Tourist Centre

Snettisham, near Hunstanton, Norfolk (0485) 543503

- ○ Open daily 10.30-5 end Mar to end Oct.
- £ Adults £3.50, children £2.50.
  Group rates: on application.
- ☞ Follow signs off A149 Snettisham bypass.
  Bus: to Snettisham village.

Facilities: 🅿 Car and coach parking 🚻 Toilets ☕ Refreshments 🎁 Gift/book shop ⛲ Play area.

Facilities for Disabled: ♿ Wheelchair access; concrete paths around farmyard for wheelchair access.

Restrictions: 🐕 Dogs on leads.

Description: Park Farm is not merely an exhibit, but is a real working farm with farmyard trails, paddocks and a sheep centre. Snettisham is home to many different types of farmyard animals and birds, including calves, pigs and turkeys. The sheep theatre demonstrates eight different breeds of sheep and shearing techniques; and a safari ride in a trailer drives through the countryside in search of deer, ospreys and wildfowl. Children can let off steam in a large play area, and learn about the animals in an education centre. Pets corner gives everyone a chance to stroke and feed the animals and in spring and early summer visitors can see lambs and deer calves being born.

On-Site Activities: Four farm trails; forty-five-minute guided tour around farm and into deer park; sheep show twice a day; wildlife quiz exhibit showing wildlife at Park Farm; pick your own eggs from 2,000 free-range hens.

Special Events: Sheep shearing on Whitsun bank holiday.

Educational Facilities: School visits.

Pensthorpe Waterfowl Park

Fakenham, Norfolk NR21 0LN Tel: (0328) 851465

- ○ Open daily 11-5 Apr to Dec; Sat and Sun 11-5 Jan to Mar.
- £ Adults £3.50, children £1.60, OAPs £3.00.
  Group rates: adults £2.50, children £1.30.
- ☞ On A1067 Norwich road 1 mile from Fakenham.
  Bus: Norwich to Fakenham bus (request stop).

Facilities: P Car and coach parking ♦♦ Toilets – mother and baby facilities ☕ Refreshments 🎁 Gift/book shop 🛝 Play area.

Facilities for Disabled: ♿ Wheelchair access. Loop induction system.

Restrictions: 🐕 Guide dogs only.

Description: Within the 200-acre reserve can be found lakeside, river, meadow and woodland trails. Hides for birdwatching enable the visitor to see wild birds which include pochards, tufted ducks, little grebes, great crested grebes, kingfishers and barn owls. The Pensthorpe collection has over 100 species to discover, including Arctic long-tailed ducks, harlequins and king eiders, as well as tiny tropical pygmy geese. Special feeding stations allow close contact with the birds. In winter, hundreds of migratory wildfowl can be seen and in the spring and summer downy ducklings, or even a kingfisher. Pensthorpe is actively involved in breeding several endangered species, one of which is the rare white-headed duck from Spain, a species that, at one point, declined to less than twenty pairs.

On-Site Activities: Free guided tours.

Special Events: Programme of special events advertised locally; send a stamped addressed envelope for full details.

Educational Facilities: Education officer/centre; school visits.

East Anglia – Norfolk 45

Strumpshaw Fen

Strumpshaw, Norwich, Norfolk Tel: (0603) 715191

○ Open daily throughout the year.
£ Admission free.
☞ Travelling east on A47 Great Yarmouth to Norwich road through Brundall, turn right after railway bridge down Low Road. Bus/Rail: Brundall station (2 miles).

Facilities: 🅿 Car and coach parking 🚻 Toilets.
Facilities for Disabled: ♿ Limited wheelchair access.
Restrictions: 🐕 No dogs.
Description: This reserve covers 800 acres of waterways, reedbeds, hay meadows, woodland and wild fowl marshes, and provides visitors with four miles of trails and four hides from which to view a wide range of birds. In summer they include marsh harriers, grebes, waders and kingfishers. In winter there are hen harriers, bean geese and whitethroats. There are many interesting plants and animals to see, including swallowtail butterflies in June.
On-Site Activities: Only by special arrangement.
Educational Facilities: Education officer; school visits.

Thrigby Hall Wildlife Gardens

Thrigby Hall, Filby, Great Yarmouth, Norfolk NR29 3DR Tel: (0493) 369477

- ○ Open daily 10-6 Easter to end Oct; 10-4 Nov to Apr.
- £ Adults £3.80, children (4-14) £2.20, OAPs £3.30. Group rates (20 or over): adults £3.30, children £2.00, OAPs £2.80.
- ☞ From Filby on A1064 Caister Road. Bus: Norwich bus from Great Yarmouth.

Facilities: 🅿 Car and coach parking 🚻 Toilets - mother and baby facilities ☕ Refreshments 🛍 Gift/book shop 🎠 Play area.

Facilities for Disabled: ♿ Wheelchair access.

Restrictions: 🐕 Guide dogs only.

Description: The one and a quarter-mile route through the landscaped grounds of Thrigby Hall provides a marvellous setting for viewing the extensive collection of animals from Asia. Sumatran tigers can be seen from the trees. There are snow leopards, gibbons, Indian porcupines, Japanese giant salamanders, green peafowl, Korean chipmunks, Sulewese macaques, Szechwan white-eared pheasants, Russian hairy-footed hamsters, Chinese alligators and many more. The park has an award-winning swamp house full of crocodiles that is well worth a visit. A unique willow pattern garden with twelve bridges and a look-out over the lake is the perfect place to relax, while for the energetic there are always the Tarzan nets or the tree walk.

On-Site Activities: Talks by prior arrangement.

Educational Facilities: Part-time education officer; school visits.

Titchwell Marsh RSPB Nature Reserve

Titchwell, King's Lynn, Norfolk, PE31 8BB Tel: (0485) 210779

○ Reserve open daily dawn to dusk throughout the year. Centre open daily 10-5 Apr to Oct, 10-4 Nov to Mar. Closed Christmas Day.

£ £2.00 per car. RSPB and YOC members free of charge.

☞ On A149 coast road 6 miles east of Hunstanton. Bus: to Hunstanton then taxi to reserve.

Facilities: 🅿 Car parking 🚻 Toilets ☕ Refreshments 🛍 Gift/book shop. Coach parking must book 6-12 months in advance.

Facilities for Disabled: ♿ Wheelchair access.

Restrictions: Restricted to public footpaths.

Description: Part of the Norfolk coastal plain, Titchwell Marsh forms an unbroken chain of natural and semi-natural habitats. This RSPB reserve features three hides which allow visitors to look over a variety of habitats including reedbeds which supply swallows with useful foods and a safe haven in which to nest, saltmarshes which provide a winter dwelling for geese, skylarks, finches and twites, and nesting sites for shelducks, mallards, partridges and buntings and freshwater marshlands which are home to over thirteen species of wildfowl, and where waders trawl the mud when the water levels are low in the autumn. One end of the beach widens into an open area of sand and shingle where terns, oystercatchers and common plovers nest in summer. The marsh provides an excellent opportunity to see birds in their natural habitat, and is recommended to all bird lovers.

On-Site Activities: Guided walks on Tuesdays, Thursdays, Saturdays and Sundays between July and August; weekends only in April, May, June, September and October.

Educational Facilities: Education officer; school visits. School visits are formal days aimed at satisfying the demands of the national curriculum.

Toad Hole Cottage and Wildlife Water Trail

How Hill, Ludham, Great Yarmouth, Norfolk NR29 5PG   Tel: (0692) 62763

- ○   Toad Hole Cottage open daily 10-5 Jun to Sep; 11-5 school holidays and local half terms; weekends and bank holidays Easter to May, and in Oct.
    Water trail open daily 10-5 Jun to Sep; weekends, bank holidays and local half terms 11-3 Easter to May, and Oct.
- £   Toad Hole Cottage admission free. Water trail: adults £2.50, children £1.50, family £5.00.
- ☞   Follow signs from A1062 Wroxham to Potter Heigham road. Bus: to Ludham. Rail: Wroxham station (5 miles).

Facilities: 🅿 Car and coach parking   🚻 Toilets   🛍 Gift/book shop   🛝 Play area.

Restrictions: 🚷 No dogs.

Description: An authentic Victorian eel-catcher's cottage set beside the River Ant has been preserved to encapture the life of a marsh family about a hundred years ago. Walking trails leading to two hides enable visitors to view the varied wildlife of the Broads. A wildlife water trail can be followed by boat (Edwardian-style) through the mysterious dykes of the How Hill Nature Reserve. The water trip lasts about fifty minutes and includes a short walk to a bird hide. Fenland (reed and sedge beds), Carr woodland and grazing marshes provide a home for yellow and white water lilies, numerous wild flowers, water birds, cormorants, marsh harriers, swallowtail butterflies and dragonflies.

On-Site Activities: The water trail is by boat with a guide/driver.

Educational Facilities: School visits. For information about educational visits, contact How Hill Trust, Ludham, Great Yarmouth, Norfolk NR29 5PG (0692) 62555.

SUFFOLK
Easton Farm Park

Easton, near Wickham Market, Woodbridge, Suffolk IP13 0EQ
Tel: (0728) 746475

- ○ Open daily 10.30-6 end Mar to end Sep.
- £ Adults £3.70, children £2.00, OAPs £3.00.
  Group rates: adults £3.00, children £1.60.
- ☞ Off A12 Wickham Market on A1120 to Earlsham.
  Bus: to Wickham Market (3 miles); Sunday bus service May to Sep.

Facilities: 🅿 Car and coach parking 🚻 Toilets – disabled and mother and baby room 🍴 Refreshments 👍 Gift/book shop 🎪 Play area.

Facilities for Disabled: ♿ Wheelchair access; hard flat pathways. Toilets.

Restrictions: 🐕 Dogs on leads.

Description: Many rare breeds of farm animals graze in the meadows here which stretch right down to the River Deben, including the rare white park cattle, longhorn cattle and red poll cattle. Suffolk horses are bred on the farm and Gloucester Old Spot pigs are kept in the yard as well as out in the fields. There is a modern dairy centre where cows are milked every afternoon. The pets paddock is a good place for children to touch and feed pygmy goats and other young farm animals. Visitors can follow a forty-minute green trail around the park over meadows, through a woodland area and out along the river bank.

    The woodland area offers good sightings of long-tailed tits, cuckoos and sometimes woodpeckers and owls, while the river bank regularly attracts moorhens and even kingfishers.

On-Site Activities: Cows milked every afternoon. Green trail walk beside River Deben.

Special Events: Held regularly throughout the season.

Educational Facilities: Education officer on request; school visits. Educational material available on request.

Kentwell Hall

Long Melford, Sudbury, Suffolk CO10 9BA Tel: (0787) 310207

- ○ Open daily spring to autumn; Sun only in winter.
- £ Adults £4.00, children £2.50, OAPs £3.40.
- ☞ On A134 4 miles north of Sudbury. Bus: to Long Melford. Rail: Sudbury station or Bury St Edmunds station (10 miles) then bus or taxi.

Facilities: ▣ Car and coach parking ♛ Toilets
  ⚏ Refreshments ♠ Gift/book shop.

Facilities for Disabled: ♿ Wheelchair access.
  Toilets for disabled currently under construction.

Restrictions: ✘ No dogs.

Description: The historic Elizabethan manor house and adjacent working farm are set amidst open parkland and are home to a number of domestic rare breeds. The farm can boast the largest flock of Norfolk horn sheep in the country, the rarest breed of sheep in England. Lincoln longwool sheep, Jersey and British white cows, Tamworth pigs and Old English goats are amongst other domestic rare breeds on the farm. Horses such as the Shire cross and the Suffolk Punch can be seen as well as a pony drawing a trap. Domestic fowl include greylag geese, grey Dorking chickens and peafowl. Carp, which were once an essential element of a countryman's diet, breed in the moat. Wool-making and Tudor-style farming methods are explained in detail.

On-Site Activities: All sorts of talks and walks available by arrangement.

Educational Facilities: School visits. Tours can be tailored to a school's particular need or interest.

Minsmere RSPB Reserve

Westleton, Saxmundham, Suffolk IP17 3BY Tel: (0728) 73281

- ○ Open Wed to Mon 9-9 or dusk throughout the year except Christmas Day and Boxing Day. Reception and shop open 9-5 Apr to Oct, 9-4 in winter.
- £ Adults £3.00, children £1.50, OAPs £2.00, RSPB and YOC members free. Group rates: groups of 10 or more must book in advance by writing to the warden (enclose SAE).
- ☞ Through Eastbridge (no coaches) from the B1122 Leiston to Yoxford road or from Westleton following signs. Rail: Saxmundham station (6 miles) and taxi.

Facilities: 🅿 Car parking 🚻 Toilets - disabled facilities 📖 Gift/book shop.

Facilities for Disabled: ♿ Wheelchair access to shop and some hides. Toilets.

Restrictions: Car drivers please take special care on narrow lanes leading to and on the reserve.

Description: Lying on the low Suffolk coast, this premier RSPB reserve offers birdwatchers an exciting experience at any time of the year. The famous Scrape – an area of shallow, brackish water, mud and islands inside the shingle beach – as well as heathland, deciduous woodland and reedbeds complete the unique habitat. A large variety of breeding birds can be seen including Britain's largest avocet colony, little terns on the beach, bitterns, water rails, bearded tits and marsh harriers in the reeds. Stonechats and nightjars may be found on the heath, nightingales and redstarts in the woods. Many waders, such as redshanks, black-tailed godwits and little stints, can be spotted. Other wildlife frequently uses the marshes. These include red deer, muntjac deer, otters and water voles, whilst on the heath you may encounter butterflies and adders.

On-Site Activities: Regular programme of guided walks. Telephone or write to warden for details.

Educational Facilities: Education officer; school visits. On-site teacher available.

The National Stud

Newmarket, Suffolk CB8 0XE Tel: (0638) 663464

- ○ Open by appointment Mar to Sep. Tours at 11.15 and 2.30 weekdays; 11.15 Sat and 2.30 Sun.
- £ Adults £3.00, children £2.00. Group rates (20 or over): on application.
- ☞ Cambridge side of Newmarket at junction of A1304, M11 (London) and A1303/A45 (Cambridge). Bus: limited bus service from Newmarket; telephone for information. Rail: Newmarket station (2 miles).

Facilities: P Car and coach parking ♦♦ Toilets ☕ Refreshments 🛍 Gift/book shop.

Facilities for Disabled: ♿ Wheelchair access.
Trail for visually handicapped by special arrangement.

Restrictions: ✘ No dogs.

Description: Originally located in Ireland, the National Stud moved to its present location on 500 acres of land leased from the Jockey Club in 1963. It is one of the principal stallion stations in Britain, with six top-class stallions. It also serves the bloodstock industry by running training courses for stud staff. A visit to the Stud takes the form of a seventy-five-minute guided tour, giving an insight into the aims and operations of a stud farm. The tour includes the National Stud stallions, the stallion unit, the foaling unit (with a short video of a foal being born) and a group of mares and foals. The Stud has been associated with many top-class racehorses including Mill Reef, the Champion European Racehorse of 1971 and the Champion Sire of 1978 and 1987. Your visit can be combined with racing in Newmarket or a visit to the National Horseracing Museum.

On-Site Activities: Guided tours.

Educational Facilities: School visits.

The Otter Trust

Earsham, near Bungay, Suffolk NR25 2AF Tel: (0986) 893470

- ○ Open daily 10.30-6 Easter to end Oct.
- £ Adults £3.50, children £2.00, OAPs £3.00.
  Group rates (school parties): £1.50 each.
- ☞ On A143 1 mile west of Bungay.
  Bus: Eastern Countries No. 631 to Earsham.

Facilities: 🅿 Car and coach parking 🚻 Toilets – disabled and mother and baby facilities 🍽 Refreshments 👍 Gift/book shop 🛝 Play area.

Facilities for Disabled: ♿ Wheelchair access. Toilets.

Restrictions: 🐕 Guide dogs only.

Description: Not only a breeding centre, but also an attractive countryside park, the Otter Trust has the largest collection of otters in the world and gives visitors the chance to experience conservation in action. The grounds encompass three lakes which are home to a large variety of waterfowl, and a river containing over twenty otter-breeding enclosures. There is also a heronry, and a picnic area where muntjac deer, wallabies and golden pheasants roam freely. The trust is situated in twenty-three acres of marshland. Each year, young otters are reintroduced into the wild. The trust works towards securing a long-term future for this, surely the most endearing of our native mammals.

On-Site Activities: Nature trail.

Educational Facilities: Education officer/centre; school visits.

Suffolk Wildlife & Rare Breeds Park

Kessingland, near Lowestoft, Suffolk Tel: (0502) 740291

○   Open daily throughout the year except Christmas Day and Boxing Day.

£   Adults £4.00, children £3.00.
      Group rates (15 or over): adults £3.00, children £2.00.

☞   Follow signs off Kessingland roundabout on main A12. Bus: to park every 30 minutes.

Facilities: 🅿 Car and coach parking 🚻 Toilets – mother and baby facilities ☕ Refreshments 🎁 Gift/book shop 🛝 Play area.

Facilities for Disabled: ♿ Wheelchair access.

Restrictions: 🚫 No dogs or other pets.

Description: Suffolk Wildlife Park features an extensive collection of African plains animals, from big cats such as lions to their large herbivorous prey, zebras and wildebeest. Other animals from around the world are also well represented: for example, tigers, monkeys and parrots. There are three explorers' trails which follow the footsteps of famous men. Children can touch and feed small animals in a special farmyard corner, and a woodland walk makes a relaxing alternative for adults. The less energetic can view the park from a safari road train.

On-Site Activities: Nature trail. Talks can be pre-arranged.

Educational Facilities: School visits.

EAST MIDLANDS & HEART OF ENGLAND

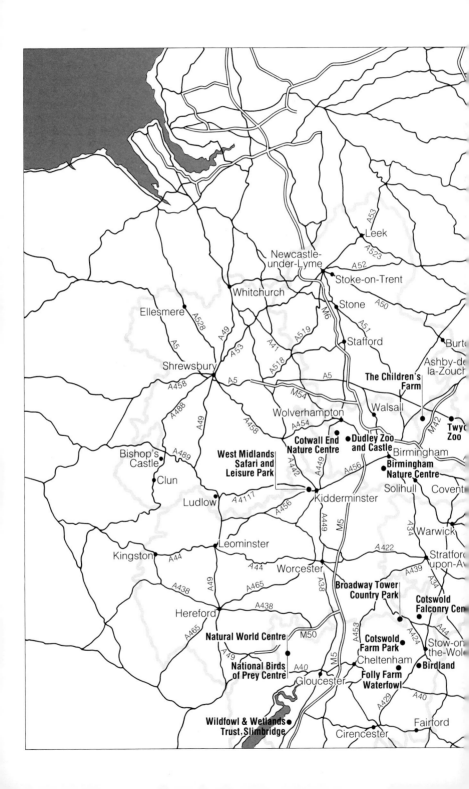

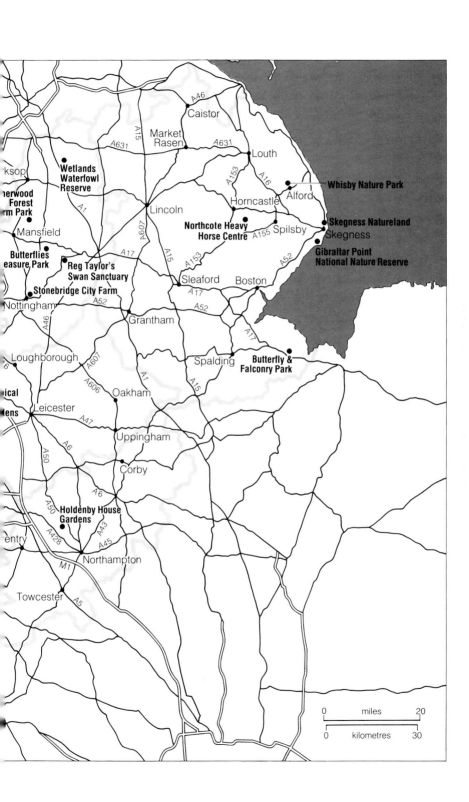

LEICESTERSHIRE
Tropical Bird Gardens

Lindridge Lane, Desford, Leicestershire LE9 9GN Tel: (0455) 824603

- ○ Open daily 10-5.30 throughout the year except Christmas Day.
- £ Adults £2.50, children £1.50.
     Group rates (over 10): 10% discount.
- ☞ M1 junction 21, B4114 to Enderby then B582 to Desford. Bus: from Leicester St Margaret bus station No. 123 to Market Bosworth (8 miles).

Facilities: 🅿 Car and coach parking 🚻 Toilets
　🍴 Refreshments 🛍 Gift/book shop.

Facilities for Disabled: ♿ Wheelchair access.

Restrictions: 🐕 No dogs.

Description: Set in five and a half acres of natural woodland, the bird gardens have spacious walk-through aviaries where you can see over thirty different species of colourful free-flying birds and handle some of the more playful ones. Species to see include parrots, macaws, Amazons, parakeets, emus, waterfowl, toucans, hornbills, owls, soft-bills, lorys and lorikeets. There is also a chick room and a large koi pond. The gardens have been carefully landscaped so that they blend with the natural surroundings and show off these wonderful tropical birds to their best advantage. Many species, including some endangered ones, have bred successfully here. There are also facilities for looking after and rehabilitating every species of British wild bird. There is a large picnic area and refreshments are available.

On-Site Activities: Woodland walks.

Educational Facilities: School visits.

LINCOLNSHIRE
Butterfly & Falconry Park

Long Sutton, near Spalding, Lincolnshire PE12 9LE
Tel: (0406) 363833/363209

○ Open daily 10-6 mid-Mar to end Oct.

£ Adults £3.20, children £2.00, OAPs £2.80.
Group rates (12 or over): adults £2.30, children £1.50.
Prices include car park and all facilities.

☞ Follow signs off A17 at Long Sutton, 14 miles north of King's Lynn.

Facilities: 🅿 Car and coach parking 🚻 Toilets – disabled and mother and baby facilities ☕ Refreshments 🛍 Gift/book shop 🛝 Play area

Facilities for Disabled: ♿ Wheelchair access. Toilets.

Restrictions: 🚫 Dogs must be kept in car park area.

Description: Set in twenty acres, the park houses one of the largest indoor tropical butterfly gardens in the British Isles, with over 500 butterflies flying freely. It is possible to observe and photograph all stages of the butterfly's life cycle among a wealth of tropical, Mediterranean and temperate flowers and foliage. There is also an insectarium which contains scorpions, giant stick insects and the notorious bird-eating spider, the tarantula.

Flying displays of eagles, hawks, owls and falcons can be seen twice daily. Visitors may handle some of the birds of prey and the park also runs courses in falconry. Other attractions include an outdoor butterfly and bee garden, wild flower meadow, wildfowl and conservation pond, farm animals and pets corner.

On-Site Activities: Guides available for butterfly house; twice-daily falconry displays.

Educational Facilities: Education officer; school visits.

Gibraltar Point National Nature Reserve

Gibraltar Point, near Skegness, Lincolnshire Tel: (0754) 762677

○   Open daily throughout the year during daylight hours.

£   Admission free, but there is a car park charge in the summer. Parties must book in advance.

☞   Follow signs from Skegness town centre to reserve 2 miles south of Skegness.
Rail: Skegness station (3 miles) then walk or taxi.

Facilities: 🅿 Car and coach parking (coaches must book in advance) 🚻 Toilets – disabled facilities 🛍 Gift/book shop.

Facilities for Disabled: ♿ Wheelchair access to some of the reserve. Toilets.

Restrictions: 🐕 Dogs on leads. No access to some shorebird breeding areas in the summer.

Description: Extending for a distance of about three miles along the Lincolnshire coast, Gibraltar Point covers over 1,000 acres of sandy and muddy seashores, sand dunes, saltmarshes and freshwater habitats. The reserve is renowned for its wealth of wildlife and provides nesting sites for numerous birds including Britain's rarest breeding seabird – the little tern. Viewpoints and hides overlook freshwater and brackish meres. Huge flocks of oystercatchers, knots and bar-tailed godwits gather during the high tides of spring and autumn, and in winter large numbers of finches and buntings frequent the marsh. A wealth of wild flowers persists year-round and at high summer the marsh is mauve with a carpet of sea lavender.

On-Site Activities: Self-guided nature trail and booklet. Walks and talks can be booked in advance. Special walks throughout the year.

Special Events: Special walks and talks throughout the year, such as wild flower walks. Events leaflet is published each year.

Educational Facilities: Education centre; school visits.

Northcote Heavy Horse Centre
Great Steeping, Spilsby, Lincolnshire PE23 5PS Tel: (0754) 86286

○ Open Sun, Wed and Thurs 11.30-4.30 Mar and Apr; Sun, Tues, Wed, Thurs 11.30-4.30 May and Jun; Sun to Fri 11.30-5.30 Jul and Aug; Sun, Wed, Thurs 11.30-4.30 Sep and Oct.

£ Adults £3.00, children £2.00, OAPs £2.50. Group rates: 10% discount.

☞ On B1195, 3 miles east of Spilsby.

Facilities: 🅿 Car and coach parking 🚻 Toilets – disabled and mother and baby facilities ☕ Refreshments 👍 Gift/book shop.

Facilities for Disabled: ♿ Wheelchair access. Toilets.

Restrictions: 🐕 No dogs.

Description: The daily programme at the centre begins with meeting the horses and explaining the history of the heavy horse. The museum takes you on a journey back in time, seeing the vehicles and equipment used while listening to what life was like a hundred years ago. The video room shows an extensive selection of subjects such as the centre's horses working around the country and the seasons shown through the eyes of the young animals growing up in the centre. Visitors can be taken for a short drive in a cart or wagon and as the horses are harnessed the onlookers are encouraged to assist. The centre also has Shetland ponies, Dales ponies and carriage horses. Rare breeds farm animals include pigs, sheep, goats, ducks and poultry. Visitors can meet representatives of all Britain's major heavy horse breeds.

Educational Facilities: Education officer/centre; school visits. For 1993, there will be a chance to dress up and experience life in Victorian times.

Skegness Natureland

North Parade, Skegness, Lincolnshire PE25 1DB Tel: (0754) 764345

- ○ Open daily from 10 throughout the year except Christmas Day, Boxing Day and New Year's Day.
- £ Adults £2.50, children £1.50, OAPs £2.00.
- ☞ At northern end of Skegness seafront. Bus/Rail: to Skegness station.

Facilities: 🅿 Council car and coach parking 100 metres away 🚻 Toilets ☕ Refreshments 🛍 Gift/book shop.

Facilities for Disabled: ♿ Wheelchair access. Ramps being installed.

Restrictions: 🐕 Dogs on leads.

Description: The baby seal rescue unit at Natureland has become known worldwide. This carefully researched rearing and treatment programme cares for abandoned seal pups that have become stranded on beaches around the Wash. The centre has dealt with many unusual animals and wherever possible returned them to the wild. Dolphins and whales, a lost walrus and a pelican, many oiled seabirds and injured birds of prey have all received care and attention at some point. Look out for the adult seals in the main seal pool. Rescued and reared from babies, these are famous for being the only 'trained' common seals in the world. The so-called training, however, is not what it seems. These seals have been encouraged to develop some play activities they instigated themselves and now produce a wide-ranging performance of tricks which they clearly enjoy. Other features of the centre include a well-stocked tropical house and aquarium, a pets corner and a huge greenhouse known as the Floral Palace. In three sections are housed the free-flight tropical butterflies and birds, flamingoes, and thousands of beautiful flowers and plants.

On-Site Activities: Informative and enjoyable talks at seals' and penguins' feeding times.

Educational Facilities: School visits. Workpacks prepared for school groups.

East Midlands – Lincolnshire   63

Whisby Nature Park

Moor Lane, Thorpe on the Hill, Lincoln Tel: (0522) 500676

○   Open daily throughout the year.

£   Admission free.

☞   Close to A46 Lincoln bypass at southern end; follow signs.

Facilities: Car and coach parking Toilets Refreshments.

Facilities for Disabled: Wheelchair access; bird hide adapted for wheelchair access.

Restrictions: Dogs on leads. Normal nature reserve rules apply as regards no swimming, cycling, horses etc.

Description: This nature park has a variety of habitats including lakes, ponds, woodland, scrub and grassland, all rich in wildlife. Magpie Walk, Coot Walk and Grebe Walk make up the six miles of nature trails from where visitors have spotted over 180 different species of bird. Even nightingales, kingfishers and sparrowhawks can be seen at the park. The lakes and ponds are the best place to look for dragonflies and damselflies. The two hides give good sightings of great crested grebes, tufted ducks and other varieties of waterfowl while many species of butterfly, including the common blue and small copper, inhabit the grassland areas.

On-Site Activities: Range of walks undertaken, usually on Sundays, covering different aspect of the nature park such as birds, flowers etc.

Special Events: About twenty events held each year covering different aspects of the wildlife of the park.

Educational Facilities: Education officer/centre; school visits.

NORTHAMPTONSHIRE
Holdenby House Gardens

Holdenby House, Holdenby, Northamptonshire NN6 8DJ Tel: (0604) 770074

- ○ Open Sun, bank holiday Mon, and Tues to Fri 2-6 Easter to end Sep.
- £ Adults £2.50, children £1.50. For house and gardens: adults £3.50, children £1.75. Group rates (20 or over): adults £2.00.
- ☞ Off A428 and A50 7 miles north-west of Northampton. About 8 miles from M1 junctions 15A and 18.

Facilities: 🅿 Car and coach parking 🚻 Toilets – disabled facilities ☕ Refreshments 🛍 Gift/book shop 🛝 Play area.

Facilities for Disabled: ♿ Wheelchair access but paths are pea gravel. Toilets.

Restrictions: 🐕 No dogs.

Description: Holdenby was once the largest house in Elizabethan England and it still has the most impressive gardens surrounding it. A large falconry centre has been established in the old Victorian kitchen garden. There are regular flying displays of different birds of prey including hawks, falcons, buzzards, kites and owls. The visitor can even book lessons in falconry. A large collection of rare breeds of farm animals lives in the gardens, including the rarest and most ancient breed of cow in Britain, the white park cow. There are Hebridean and Soay sheep as well as Gloucester Old Spot pigs and the Vietnamese pot-bellied pigs. Pygmy goats wander around the gardens and visitors can look out for the four rare breeds of pheasant: the Lady Amhurst, silver, golden and yellow golden. There is a cuddle farm where children can stroke or touch many baby animals including piglets, lambs and ducklings.

On-Site Activities: Falconry courses and guides by appointment only. Cuddle farm.

Special Events: Programme on application.

Educational Facilities: Education officer/centre; school visits.

NOTTINGHAMSHIRE
Butterflies Pleasure Park

The White Post, Farnsfield, Nottinghamshire NG22 8HX Tel: (0623) 882773

○ Open daily 10-6 or dusk throughout the year.

£ Adults and children £2.00. Group rates: £1.50.

☞ On A614 12 miles north of Nottingham at Rainworth-Southwell crossroads. Bus: from Mansfield to Newark and Nottingham to Doncaster, Retford or Worksop. Also the Sherwood Forester.

Facilities: 🅿 Car and coach parking 🚻 Toilets ☕ Refreshments 🛍 Gift/book shop 🛝 Play area.

Facilities for Disabled: ♿ Wheelchair access.

Restrictions: 🐕 Dogs allowed to exercise in 5-acre area.

Description: The large tropical glasshouse is home to around twenty different species of butterfly which fly freely through this simulated rainforest with waterfalls and ponds. Golden finches and weaver birds are also free-flying, while carp swim in the pools. A garter snake and red-kneed tarantula are also displayed.

Outside, a guided nature trail introduces the visitor to a number of native plants and animals. This educative experience takes about half an hour. The trail leads through woodland and in spring wild flower meadows appear. A play area is provided for children, as well as tractor rides.

On-Site Activities: Guides, walks, talks, worksheets. Tropical house with 350 tropical butterflies plus plants, birds and fish.

Educational Facilities: School visits.

Reg Taylor's Swan Sanctuary

Hill Farm Nurseries, Normanton, Southwell, Nottinghamshire NG25 0PR
Tel: (0636) 813184

- ○ Open daily 10-5.30 throughout the year.
- £ Admission free.
- ☞ East of Southwell on Hockerton road.
  Bus: to Southwell (10 minutes walk).

Facilities: Car and coach parking Toilets Refreshments Gift/book shop.

Facilities for Disabled: Wheelchair access.

Restrictions: Dogs on leads.

Description: Naturally landscaped grounds of nine and a half acres provide an ideal setting in which to see not only the sanctuary's swans but also a wealth of other bird life. There are large numbers of mallard and Canada geese and the four lakes within the grounds also support more unusual water birds such as the heron, kingfisher and moorhen. The centre cares for injured swans and other birds, many having been rescued by the RSPCA and brought to the sanctuary. Although for some it becomes a safe haven for life, for many of the birds it is a resting place to recuperate and they will fly free once their health and strength have returned. During the winter months in particular, the odd special visitor returns and on occasions there is the added pleasure of a new mate or family accompanying them to take advantage of the food on offer.

On-Site Activities: Organised walks and talks for schools and others by appointment. Birds can be fed by visitors.

Special Events: Nottinghamshire Wildlife Trust events at weekends. Nottinghamshire Hardy Plant Society summer show.

Educational Activities: School visits.

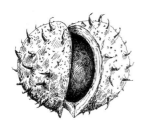

Sherwood Forest Farm Park

Lamb Pens Farm, Edwinstowe, Mansfield, Nottinghamshire NG21 9HL
Tel: (0623) 823558

- ○ Open daily 10.30-5.30 Easter to mid-Oct. Pre-booked groups may arrive from 9.30.
- £ Adults £2.50, children £1.75, OAPs £2.25. Group rates: school groups, adults and children £1.25.
- ☞ Just off A6075 between Edwinstowe and Mansfield Woodhouse. Bus: to Warsop Windmill (10 minutes walk).

Facilities: ⓟ Car and coach parking ♦♦ Toilets – mother and baby facilities ☕ Refreshments 🎁 Gift/book shop 🛝 Play area.

Facilities for Disabled: ♿ Wheelchair access.

Restrictions: 🐕 No dogs; kennels are available free of charge.

Description: The park is set in a secluded valley on the edge of Sherwood Forest and is attractively landscaped with two lakes catering for an interesting collection of wildfowl, ducks, swans and geese. There is a large collection of rare breeds – protected species of cattle, sheep, pigs, goats and horses – and a large display of poultry in grassy fields alongside the waterfowl. The extensive water gardens and aviary of exotic, colourful birds are also a popular feature. Other facilities include a pets corner, safe play areas and picnic sites. The Rare Breeds Survival Trust is an organisation set up to promote the conservation and study of endangered breeds of British farm animals and the centre at Sherwood Forest is an officially approved farm park.

On-Site Activities: Tractor and trailer rides (subject to demand).

Educational Facilities: Education officer; school visits.

Stonebridge City Farm

Stonebridge Road, St Anns, Nottingham NG3 2FR Tel: (0602) 505113

- ○ Open daily 9-5 throughout the year.
- £ Admission free; donations welcome.
- ☞ Follow signs from city centre to Sneinton. At crossroads of Bath Street and Carlton Road, turn left, second right, second right. Bus: Nos. 23, 24, 25, 26, 27, 29, 64, 65, 66 from city centre.

Facilities: ⓟ Car and coach parking ⚦ Toilets – mother and baby facilities ☕ Refreshments ⚠ Play area.

Facilities for Disabled: ♿ Wheelchair access.

Restrictions: 🐕 No dogs.

Description: Established in 1978 by a residents' committee, this inner city farm covers three acres, and apart from farm animals such as pigs, goats, sheep, geese, ducks and chickens, the pets corner contains rabbits and guinea pigs. The site has been developed to include an allotment field, a farm garden, paddocks, a wildlife area for plants, a barn and a play area. There are daily milking demonstrations and seasonal attractions such as sheep-shearing demonstrations and lambing. The farm encourages all visitors to gain experience of organic farming by inviting them to become volunteers. Wellies, gardening gloves and tools are all freely available for a one-off experience or a regular commitment!

On-Site Activities: Guided tours (pre-booked for large parties); displays and information board.

Special Events: Cheese-making days; summer play schemes; volunteers' club; monthly kids' club.

Educational Facilities: Education officer/centre; school visits. Inter-schools competitions; hands-on experience for those taking pre-vocational certificates; loan of incubator and brooder.

East Midlands – Nottinghamshire 69

Wetlands Waterfowl Reserve

Lound Low Road, Sutton cum Lound, Retford, Nottinghamshire
Tel: (0777) 818099

- ○ Open daily 10-5.30 or dusk throughout the year.
- £ Adults £1.50, children £1.00.
  Group rates: on application.
- ☞ Off A638 Retford to Bawtry road.
  Bus: Bawtry bus from Retford or the Sherwood Forester.

Facilities: 🅿 Car and coach parking 🚻 Toilets
🍽 Refreshments 🛍 Gift/book shop ⛰ Play area.

Facilities for Disabled: ♿ Wheelchair access.

Restrictions: 🐕 No dogs.

Description: This reserve covers over thirty acres and has over 100 species of waterfowl, geese, swans and ducks on two lagoons. There is also an aviary with parrots, and a bird sanctuary and rescue centre which has a large number of tawny owls, along with injured buzzards and kestrels. On the children's farm there are rare breeds of sheep, pygmy goats and wallabies. Visitors can take advantage of trails, hides and fishing facilities.

On-Site Activities: Guide for parties of over twenty if pre-booked.

Educational Facilities: School visits.

GLOUCESTERSHIRE
Birdland

Rissington Road, Bourton-on-the-Water, Gloucestershire GL54 2BN
Tel: (0451) 820480/820689

- ○ Open daily 10-6 Apr to Oct, 10-4 Nov to Mar.
- £ Adults £3.00, children and OAPs £2.50. Group rates (over 10): 10% discount.
- ☞ Rail: Moreton-in-Marsh station (4 miles).

Facilities: Parking in adjacent car park. Toilets in adjacent car park. Refreshments. Gift/book shop.

Facilities for Disabled: Wheelchair access.

Restrictions: Dogs on leads.

Description: Situated on the banks of the River Windrush near to the picturesque Cotswold village of Bourton-on-the-Water, Birdland enjoys the reputation of having one of the finest collections of foreign birds in the country. Penguins, waterfowl, tropical and subtropical birds can be seen. A special feature of these gardens is the great variety of birds flying free, many of them strikingly coloured, tropical species. Look out for the blue and gold macaws, scarlet macaws, hyacinthine macaws, rainbow lorikeets, yellow-backed lorys, black lorys, umbrella cockatoos, red-breasted geese, night herons and many species of duck. Many of the birds bred at Birdland have been reared for the first time in captivity and the centre is actively involved in conservation work. The real stars of Birdland are the penguins. Not content just to entertain their public with impressive underwater acrobatics, some are now film stars and stole the show in the film *Batman Returns*!

Educational Facilities: School visits.

Cotswold Falconry Centre

Batsford Park, Moreton-in-Marsh, Gloucestershire GL56 9QB
Tel: (0386) 701043

○ Open daily 10.30-5.30 Mar to Oct, weekends only Nov and Dec.

£ Adults £2.50, children £1.00, concession £1.50. Group rates (over 20): 10% discount.

☞ On A44 1 mile west of Moreton-in-Marsh. Rail: Moreton-in-Marsh station (1½ miles).

Facilities: 🅿 Car and coach parking 🚻 Toilets 🍽 Refreshments 🎁 Gift/book shop 🎠 Play area.

Facilities for Disabled: ♿ Wheelchair access.

Restrictions: 🐕 No dogs.

Description: The centre is part of Batsford Park and is located adjacent to the Batsford Arboretum. It offers flying displays of eagles, owls, hawks and falcons, and not only is there a chance to appreciate their speed, grace and agility, but also an opportunity to handle some of these magnificent birds. The collection ranges from the golden eagle to the little owl, and many other owl species can be observed in their natural habitat in the Owl Wood. The centre specialises in breeding and conservation and has many breeding pairs. Young birds can be seen for most of the year. The fifty-acre arboretum contains 1,200 species of trees, many rare, with superb views over the vale of Evenlode.

On-Site Activities: Guided tours for pre-booked parties. Flying displays throughout the day.

Educational Facilities: Education officer/centre; school visits.

Cotswold Farm Park

Guiting Power, Cheltenham, Gloucestershire GL54 5UG Tel: (0451) 850307

- ○ Open daily 10.30-6 Apr to end Sep.
- £ Adults £3.00, children £1.50. Group rates: 10% discount.
- ☞ M5 junction 9 then off B4077 Stow-on-the-Wold to Tewkesbury road. Rail: Moreton-in-Marsh (8 miles).

Facilities: Car and coach parking Toilets Refreshments Gift/book shop Play area.

Facilities for Disabled: Wheelchair access.

Restrictions: No dogs in animal exhibition.

Description: This is a typical Cotswold farm, situated in the centre of the Cotswold Area of Outstanding Natural Beauty. Perched on top of a hill, the park has the most comprehensive collection of rare breeds of British farm animals in the country. Its 1,173 acres is made up of 884 cereal, 125 permanent grass, 111 rotation grass, 43 farm park and 10 acres of buildings and roads. The livestock includes Soay sheep, longhorn cattle, Gloucester Old Spot pigs, Shire horses and Bagot goats. The farm trail has been designed to show a modern sheep and arable farm, and not only are the growing crops described but also features such as hedgerows, previously quarried areas and other aspects of the landscape. The overall management of the farm has maintained the critical balance of nature with the economic requirements of the present day. The park has many seasonal activities including sheep shearing and lambing; as well as regular features such as milking and a pets corner where children may feed lambs, goat kids and calves.

On-Site Activities: Lambing, shearing and other seasonal exhibitions.

Education Facilities: Education officer/centre; school visits. Lambing (1,000 ewes) in late September, horse shoeing and other seasonal farming exhibitions.

Folly Farm Waterfowl

Folly Farm, near Bourton-on-the-Water, Cheltenham, Gloucestershire GL54 3BY Tel: (0451) 820285

○ Open 10-6 Apr to Sep, 10-4 or dusk Oct to Mar. Closed Christmas Day and Boxing Day.

£ Adults £2.70, children £1.60, OAPs £2.00. Group rates (20 or over): 10% discount for booked groups.

☞ On south side of A436 Bourton to Cheltenham Road, 2½ miles from Bourton. Bus: Pulhams buses from Cheltenham or Bourton.

Facilities: ⓟ Car and coach parking ♦♦ Toilets – disabled facilities ☕ Refreshments in summer 🛍 Gift/book shop and garden centre

Facilities for Disabled: ♿ Wheelchair access. Toilets.

Restrictions: 🐕 Dogs on leads.

Description: The farm is set in the heart of the Cotswold countryside and is spaciously laid out for watching over 120 different breeds of domestic waterfowl and wildfowl. The aim of the farm has always been to conserve and popularise pure breeds. There is every sort of duck from the Indian runner to the miniature silver appleyard duck. A large number of different pure breeds of geese can be found in their various pens, including the ornamental Chinese goose, which has many swan-like characteristics, and the Sebastopol goose, which looks more like the goose of the traditional pantomime. Over fifty different species of wild waterfowl live happily together on large pools that flow into two adjoining lakes. Children will enjoy the undercover pets area where guinea pigs, rabbits, goats and deer have been hand-reared and so will eat from their hands.

On-Site Activities: Informal talks on request.

Educational Facilities: School visits.

National Birds of Prey Centre

Newent, Gloucestershire GL18 1JJ Tel: (0531) 820286

- ○ Open daily 10.30-5.30 Feb to end Nov.
- £ Adults £3.75, children £1.95. Group rates: (adult parties) adults £2.25, children £1.75; (children's parties) adults £1.95, children £1.30.
- ☞ Leave Gloucester on A40 then turn right on to B4215 to Newent then follow signs. Bus: Gloucester to Newent.

Facilities: 🅿 Car and coach parking 🚻 Toilets – disabled facilities ☕ Refreshments 🛍 Gift/book shop ⛹ Play area.

Facilities for Disabled: ♿ Wheelchair access.

Restrictions: 🚫 No dogs.

Description: The centre has grown from a private collection of twelve birds in 1967 to one of the largest collections of birds of prey in the world and second largest in Europe, after Berlin Zoo. The twelve acres of rolling Gloucestershire countryside, with views of the Cotswolds and Malvern Hills, are home to some 200 birds: hawks, falcons, owls, eagles, vultures, condors and a very friendly secretary bird. It is an ideal spot for photography as many birds are uncaged. Regardless of the number of visitors, there are four flying displays a day, weather permitting, each lasting forty-five minutes. These are all specifically timed so visitors do not miss a 'fly past'. A minimum of three birds fly in any one display and these are selected from the eagles, vultures, hawks, falcons and owls. The hawk walk also gives unobstructed views of a wide variety of the trained birds on blocks – ideal for photography.

On-Site Activities: Flying demonstrations four times a day; guided tours for booked parties.

Educational Facilities: Education officer/centre; school visits. Guided tours for school parties. Talks (with birds) in schools. Education hall offers space for related projects and activities.

Natural World Centre

Springbank, Birches Lane, Newent, Gloucestershire GL18 1DN
Tel: (0531) 821800

○ Open daily 10-5 Easter to end Oct, weekends only in winter.

£ Adults £2.95, children £1.95, family £8.95.
Group rates: adults £2.00, children £1.30.

☞ From B4221 from Ross-on-Wye or B4215 from Gloucester, take B4215 to Leominster and follow signs 1¾ miles from junction. Bus: from Newent town, take Newent to Dymock bus.

Facilities: ▣ Car and coach parking ♦♦ Toilets ☕ Cold drinks and confectionery ♣ Gift/book shop ⚠ Play area.

Facilities for Disabled: ♿ Wheelchair access.

Restrictions: ✖ No dogs.

Description: The centre is situated in the Gloucestershire countryside close to the borders of Hereford and Worcester with views over the Malvern Hills. It features a tropical jungle-style garden where butterflies from South America, the Philippines and Malaysia fly freely amongst the flowers. A wide variety of reptiles is on display, including some quite rare species such as the Australian bearded dragon. There is also a large collection of invertebrates featuring some twenty-four types of tarantula. There are chipmunks, rabbits, guinea pigs, rare fowl and waterfowl. The aviaries house parakeets, lovebirds, finches, canaries, doves and many more birds. The large static nature exhibition displays include butterflies, birds, mammals and insects. Also to be seen are many wasps and bees of the world and the exploded pellet of a barn owl containing the bony remains of its last catch.

On-Site Activities: Informative talk on butterflies and display and touching of pythons with detailed talk arranged for group visits. Talks can be arranged or questions answered on any aspect of natural history.

Educational Facilities: School visits (see above).

The Wildfowl & Wetlands Trust, Slimbridge

Slimbridge, Gloucestershire GL2 7BT   Tel: (0453) 890065

- ○ Open daily 9.30-5 in summer, 9.30-4 in winter throughout the year except Christmas Eve and Christmas Day.
- £ Adults £4.25, children £2.15, OAPs £3.20, family £10.00. Group rates: on application.
- ☞ M5 junctions 13 or 14 then follow signs. Bus: difficult; please telephone centre for advice.

Facilities: 🅿 Car and coach parking 🚻 Toilets – mother and baby facilities ☕ Refreshments 🛍 Gift/book shop.

Facilities for Disabled: ♿ Wheelchair access. Trails for visually handicapped.

Restrictions: 🐕 No dogs.

Description: Slimbridge is reputed to have one of the largest collections of waterfowl in the world, displayed in as close to a natural environment as possible. These include all six species of flamingo in recreated lakes and a variety of geese and ducks in a wetland setting. In the tropical house, hummingbirds hover. In addition, there are large, heated birdwatching hides enabling the public to enjoy the spectacular winter sights of thousands of migrating birds which take refuge on the reserve. After their 2,500-mile flight from Siberia, Bewick swans make Slimbridge their wintering grounds. The major conservation achievement of the collection was to bring back from the edge of extinction the Hawaiian goose, which is now common in the collection. Slimbridge actively encourages visitors to feed the birds and many bold geese will actually feed from the palm of your hand.

On-Site Activities: Full range of activities available.

Special Events: Calendar of events available on request.

Educational Facilities: Education officer/centre; school visits.

HEREFORD AND WORCESTER
Broadway Tower Country Park

Broadway, Worcestershire WR12 7LB Tel: (0386) 852390

○ Open daily 10-6 Apr to end Oct; last admissions at 5.15.

£ Adults £2.50, children £1.50, family £7.00, concession £1.50. Group rates: adults £1.50, children £1.00.

☞ Follow signs off A44 Evesham to Oxford road.

Facilities: 🅿 Car and coach parking ♦♦ Toilets ⏣ Refreshments 🛍 Gift/book shop 🅼 Play area.

Facilities for Disabled: ♿ Limited wheelchair access.

Restrictions: 🐕 Dogs on leads.

Description: The park is privately owned and managed and offers exhibitions, woodland walks and a children's farm. The animal encounters include red deer, pigs, donkeys and rabbits, as well as special breeds such as the Broadway-Will herd of pygmy goats, Cotswold sheep and the magnificent Highland cattle. The centrepiece of the park is Broadway Tower, a folly built in the late 1700s by the Earl of Coventry which offers views over twelve counties and houses display cabinets of many wild mammals and birds. There are two suggested walks (one and a half miles and two miles) which guide visitors around the most interesting parts of the rich landscape outside the park boundary. Within the park are picnic and barbecue areas, an adventure playground and a giant chess and draughts board.

Educational Facilities: Country classroom; school visits.

West Midlands Safari and Leisure Park

Spring Grove, Bewdley, Worcestershire DY12 1LF Tel: (0299) 402114

- ○ Open daily 10-5 weekends, 10-4 weekdays Apr to end Oct.
- £ Adults and children £3.75 (rides extra). Group rates: £2.75.
- ☞ A456 from Birmingham. Situated between Kidderminster and Bewdley.

Facilities: 🅿 Car and coach parking 🚻 Toilets – mother and baby facilities ☕ Refreshments 🛍 Gift shop 🛝 Play area.

Facilities for Disabled: ♿ Wheelchair access.

Restrictions: 🐕 No dogs; kennels provided.

Description: The safari park has three main reserves – African, American and Eurasian. The African is the first and contains giraffes, zebras, ankole cattle, brindled gnus, camels and elands. Inside, lions have their own fenced enclosure and eye potential prey grazing outside. Unlike many other zoos and safari parks, visitors are allowed to feed certain animals from their cars by buying bags of approved dietary pellets. And the animals have taken full advantage, pressing their heads through car windows to pluck the pellets. In the Eurasian reserve are Barbary sheep, Przewalski's horses, deer and yaks, all eyed by tigers, once again in separate enclosures. The American reserve is home to wapiti, timber wolves and guanacos. In another part of the park are wallabies and a very audacious troop of rhesus monkeys. There is also a small zoo with a children's farmyard with llamas, ponies, pygmy goats and pot-bellied pigs, as well as a sealion display and an interesting collection of snakes. The park also has a large amusement area.

Educational Facilities: School visits.

STAFFORDSHIRE
The Children's Farm

Ash End House Farm, Middleton, near Tamworth, Staffordshire B78 2BL
Tel: (021 329) 3240

- ○ Open daily 10-6 or dusk throughout the year except Christmas Day.
- £ Adults £1.10 (half price), children £2.20 (including feed, badge etc.). Group rates: on application.
- ☞ Off A4091, follow signs.

Facilities: 🅿 Car and coach parking 🚻 Toilets – disabled facilities ☕ Refreshments 🛍 Gift/book shop 🛝 Play area.

Facilities for Disabled: ♿ Wheelchair access to most animals. Toilets.

Restrictions: 🐕 Guide dogs only.

Description: Touch, see, smell and hear all the sights and sounds of a working farm. Ash End House Farm has been set up specifically with children in mind, with guided tours and plenty of opportunities to get close to friendly animals: bottle-feeding goat kids, stroking the calves, or holding newly hatched chicks. There is a host of animals to see from the largest – a gigantic Shire horse called Henry – to the smallest ducklings. There are many rare breeds as well, including Bagot goats, saddleback pigs and Soay sheep.

On-Site Activities: Guided tours available for all pre-booked groups. Mini-tours for all visitors every hour during school holidays and at weekends.

Special Events: Wool spinning demonstrations weekly; sheep shearing in spring; Father Christmas/nativity in December.

Educational Facilities: School visits.

WARWICKSHIRE
Twycross Zoo

Atherstone, Warwickshire CV9 3PX   Tel: (0827) 880250

- ○ Open daily 10-6 Mar to Oct, 10-4 Nov to Feb. Closed Christmas Day.
- £ Adults £3.80, children (3-14) £2.00, OAPs £2.60. Group rates: on application.
- ☞ On A444 Burton to Nuneaton road directly off M42 junction 11. Coach: some coach companies arrange excursions during the summer.

Facilities: 🅿 Car and coach parking ♦♦ Toilets – disabled and mother and baby facilities ☕ Refreshments 🛍 Gift/book shop 🅰 Play area.

Facilities for Disabled: ♿ Wheelchair access; cafeterias can accommodate wheelchairs and the majority of animal houses have ramps for easy access. Guided trails for visually handicapped can be arranged.

Restrictions: ✘ No dogs.

Description: Twycross Zoo hosts Britain's largest primate collection. There are numerous vociferous and gymnastic gibbons, wonderfully gentle gorillas as well as curious chimps. Bonobos (the other species of chimp) are also found here, giving a fascinating view of our closest relative. Orang-utans complete the ape contingent, but many smaller monkeys such as marmosets and tamarins also breed here. The list of other animal residents includes elephants, giraffes, otters, lions, Sumatran tigers, birds and reptiles. This is the only place in Britain where you will find the delightful and rare Baikal seal. The sealion pool and penguin pool with underwater viewing are other areas to see aquatic life. The enchanted forest is a showplace for smaller creatures like bats and bugs. Other attractions include the Rio Grande railway, donkey rides and bouncy castle. You can meet and handle some of the animals, and watch others being fed.

On-Site Activities: Exhibitions for the whole family; animal chats and handling; sealion and penguin feeding; meet a zoologist; interactive graphics; guided tours can be booked in advance.

Educational Facilities: Education officer/centre; school visits. Lively hands-on teaching sessions linked to the national curriculum with a cross-curricular approach. Comprehensive teacher literature, courses and advice. Free preliminary visits for teachers and group leaders. Caters for all levels of primary, secondary and tertiary education.

WEST MIDLANDS
Birmingham Nature Centre

Pershore Road, Birmingham B5 7RL Tel: (021) 471 4997

- ○ Open daily 10-4.30 Mar to end Oct, weekends 10-dusk in winter.
- £ Admission free.
- ☞ On A441 Pershore Road, opposite BBC Pebble Mill. Bus: Nos. 41, 45, 47 from city centre to Pershore Road.

Facilities: Car and coach parking ♦ Toilets – mother and baby facilities ♦ Refreshments ♦ Gift/book shop.

Facilities for Disabled: Wheelchair access.

Restrictions: Guide dogs only.

Description: Two miles from the city centre, the Birmingham Nature Centre combines the attractions of a zoo, nature trail and museum, all on a six-acre site. In natural outdoor enclosures, visitors can see past and present examples of Britain's wildlife including lynx, beaver, wildcat, fox and snowy owl. Fallow deer, Soay sheep and wild rabbits graze in spacious grassy paddocks. There is also an indoor display of British rodents, from the diminutive harvest mouse to the fat or edible dormouse. There are opportunities to watch amphibians and reptiles at close quarters, and there are underwater windows to observe otters, beavers and fish. As for insects, there is an observation beehive, wood-ants' nest and butterfly aviary. A new aviary is to house a colourful selection of British finches.

On-Site Activities: Guided walks.

Special Events: Sheep shearing; harvest conservation and craft displays.

Educational Facilities: Education officer/centre; school visits. Maximum 120 children at any one time; visits should be booked in advance. Class sessions at study base, teachers' notes on request.

Cotwall End Nature Centre

Catholic Lane, Sedgley, Dudley, West Midlands DY8 4QN
Tel: (0902) 674668

○ Open daily 9-7 Apr to Sep, 9-4.30 Oct to Mar.

£ Admission free.

☞ Follow signs off Wombourne to Sedgley road and Dudley to Wolverhampton road. Bus: No. 588 Dudley to Wolverhampton to top of Catholic Lane.

Facilities: 🅿 Car and coach parking 🚻 Toilets – mother and baby facilities ☕ Refreshments 🎁 Gift/book shop 🛝 Play area.

Facilities for Disabled: ♿ Wheelchair access.

Restrictions: 🐕 Dogs on leads.

Description: This centre comprises a collection of native wild animals which includes the daily flying of owls, falcons and hawks; a collection of domesticated animals – pigs, sheep, goats and bantams, many of which can be hand-fed – and some pets. There are also 150 acres of ancient woodland, ponds, streams and meadows laid out with nature trails. Wildlife that can often be seen includes marsh orchids, kingfishers, kestrels, up to twenty species of butterfly, breeding sparrowhawks and green and great spotted woodpeckers. There are also craft studios.

On-Site Activities: Planned activities for school and other groups.

Special Events: Country day on first Sunday in September as well as other weekend events.

Educational Facilities: Education officer/centre; school visits.

Dudley Zoo and Castle

2 The Broadway, Dudley, West Midlands DY1 4QB Tel: (0384) 252401

○ Open daily 10-4.30 in summer, 10-3 in winter.

£ Adults £4.00, children £2.00, family £11.50. Group rates: adults £3.10, children £1.75; (schools) adults £2.60, children £1.45.

☞ M5 junction 2. Bus: to Dudley. Rail: Dudley Port station (2½ miles).

Facilities: 🅿 Car and coach parking 🚻 Toilets – mother and baby facilities 🍽 Refreshments 🛍 Gift/book shop 🛝 Play area.

Facilities for Disabled: ♿ Wheelchair access.

Restrictions: 🐕 Guide and listening dogs only.

Description: Dudley Zoo occupies the slopes of a high, round hill which is topped by the ruins of a thirteenth-century castle. By following the long, spiral route around and up the hill, visitors are taken past flamingoes, aviaries and a large enclosure containing spotted hyenas. In a terraced paddock, there is a herd of Barbary sheep, and other enclosures on this steep route hold Arabian gazelles – an endangered species which the zoo was the first to keep – fallow deer and African elephants. Halfway along this route is a wildfowl walk, heavily wooded with each enclosure exhibiting birds from a different continent. There is a children's corner, reptile and invertebrate houses and a farmyard. In a wonderful glass-fronted enclosure with an artificial stream are some Asian short-clawed otters. The monkey house has lar gibbons, grivet and patas monkeys and other species. The zoo is home to some 250 different species which include Himalayan bears, chimpanzees and orang-utans.

On-Site Activities: Audio-visual presentation of castle history. Guided tours of castle if pre-booked.

Special Events: Special events throughout the year, usually involving historical re-enactments. Santa's grotto at Christmas.

Educational Facilities: Education officer/centre; school visits.

LONDON

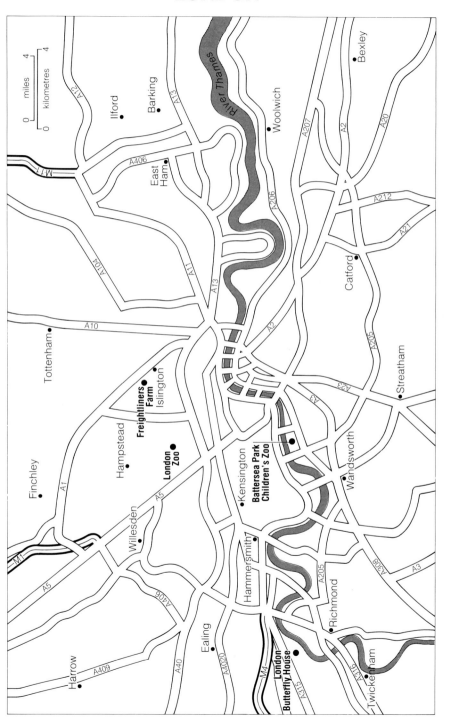

Battersea Park Children's Zoo

Wandsworth Borough Council Parks Department,
Wandsworth Town Hall, London SW18 Tel: (081) 871 6347

○ Open daily 11-6 Easter to end Sep; weekends only 10.30-3.30 Oct to Jan.

£ Adults 75p, children 25p, OAPs and disabled visitors free of charge. Group rates (booked): 20p. Season tickets: adults £7.50, children £2.50.

☞ Enter by Albert Bridge Road or Queenstown Road.
Bus: Nos. 19, 39, 44, 45, 49, 137, 170.
Rail: Battersea Park or Queenstown Road stations, Sloane Square underground (District Line).

Facilities: 🅿 Car and coach parking 🚻 Toilets
🍽 Refreshments 👍 Gift/book shop 🛝 Play area in park.

Facilities for Disabled: ♿ Wheelchair access.

Restrictions: 🐕 No dogs.

Description: Near the pagoda in Battersea Park, the zoo houses an endearing collection of animals from all over the world. Ring-tailed lemurs and other primates are situated near the entrance within a stone's throw of the gift and souvenir shop. The zoo has a collection of snakes, lizards and amphibians, as well as a wide variety of mammals. Meerkats and mongooses live in mounded burrows in their own social groups overlooking a large otter enclosure. South American rodents are represented, as well as a selection of birds including flamingoes, rheas, parrots and toucans. Children can play with farm animals in the children's area where goats, cows and donkeys roam free.

On-Site Activities: Pony rides; birthday parties can be pre-booked.

Educational Facilities: School visits.

Freightliners Farm

Sheringham Road, Islington, London N7 8PF   Tel: (071) 609 0467

- ○ Open Tues to Sun 9-1 and 2-5 throughout the year.
- £ Admission free; donations welcome.
- ☞ Near A1 Holloway Road. Bus: Nos. 14, 19, 43, 45, 104, 271 and 279. Rail: Highbury and Islington underground (Victoria Line), Caledonian Road and Holloway Road underground (Picadilly Line).

Facilities: ♦♦ Toilets – disabled and mother and baby facilities ☕ Refreshments.

Facilities for Disabled: ♿ Wheelchair access. Toilets.

Restrictions: ✗ No dogs.

Description: Situated in the heart of London, the farm provides a home to a host of farmyard animals including chickens, ducks, cows, sheep, pigs and goats. There are interesting activities all year round from sheep-shearing demonstrations, yoghurt and butter-making and honey production to pig races. Children are invited to bring along their boots and join in the farm's daily functions, such as feeding, milking and looking after the animals. Alternatively, relax in the wildlife garden, complete with frogs, newts and wild flowers. The farm is fully operational, and two workers are on duty at all times, so it is always a safe place for children to play. There are special open days twice a year.

On-Site Activities: Seasonal activities relating to a working farm: sheep shearing, butter-making, honey production etc.

Educational Facilities: School visits.

London Butterfly House

Syon Park, Brentford, Middlesex TW8 8JF Tel: (081) 560 7272

○ Open daily 10-5 in summer, 10-3.30 in winter throughout the year except Christmas Day and Boxing Day.

£ Adults £2.20, children and OAPs £1.40, family £6.50. Group rates: on application.

☞ Within Syon Park, Brentford. Signposted Syon House and Syon Park within 2-mile radius. Bus: Nos. 237 or 267 to Brent Lea (3 minutes walk). Rail: Gunnersbury underground station (District Line).

Facilities: 🅿 Car and coach parking ♦♦ Toilets
   ⏻ Refreshments 🎁 Gift/book shop 🛝 Play area.

Facilities for Disabled: ♿ Wheelchair access.

Restrictions: 🐕 No dogs.

Description: Set in the grounds of beautiful Syon Park, the butterfly house consists of a large tropical greenhouse where hundreds of species of butterfly can roam freely. Their natural warm and humid environment is recreated indoors using lights, water and heating. The butterflies come from all around the world and many are bred at Syon Park. Some species are more colourful than others, and many extremely spectacular examples can be seen. More information about butterflies can be found in the adjoining shop, and the staff are very helpful and informative. There is also an insect, amphibian and reptile exhibition.

Educational Facilities: Education officer/centre; school visits.

London Zoo

Regent's Park, London NW1 4RY Tel: (071) 722 3333

○ Open daily 10-5.30 in summer, 10-4 in winter.

£ Adults £6.00, children £3.70, concession £4.70. Group rates: on application. Car park £4.00 per day.

☞ Free parking on outer circle in Regent's Park after 11. Large car park. Bus: No. 274 from Baker Street. Rail: Camden Town underground (Northern Line).

Facilities: 🅿 Car and coach parking 🚻 Toilets – mother and baby facilities ☕ Refreshments 🛍 Gift/book shop 🅰 Play area.

Facilities for Disabled: ♿ Wheelchair access.

Restrictions: ✘ No dogs.

Description: Home to over 8,000 animals, the zoo is also one of the largest contributors to conservation programmes in the world. In the small mammal pavilion, endangered species like the golden lion tamarin rub shoulders with more familiar animals such as the black rat. Underneath, a stairway leads visitors to the moonlit world of nocturnal animals: echidnas, aardvarks, foxes and bats. The invertebrate house contains tarantulas, jellyfish, giant snails and stick insects; all safely behind glass! The ungulate section has a substantial collection of hoofed animals, including giraffes and the rare Arabian oryx, which was extinct in the wild but has been saved. Many other rare species are bred and some are on display, notably the world's rarest lions, giant pandas, Sumatran tigers and black rhinos. Children are catered for in a discovery centre where they can pretend to be a giraffe or discover why camels need big feet. Animals in Action shows take place daily, and animal encounter sessions allow children to come face to face with unusual beasts. Not to be missed are the reptile house and the aquarium, which has one of the best collections of fish in the country.

On-Site Activities: Animals in Action shows; animal encounters, feeding times.

Educational Facilities: Education officer/centre; school visits.

NORTHUMBRIA

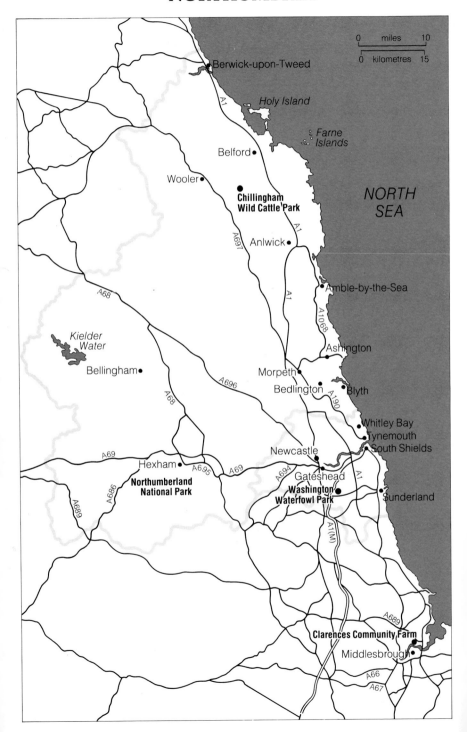

CLEVELAND
Clarences Community Farm

Behind Port Clarence Road, Port Clarence, Middlesbrough, Cleveland
Tel: (0642) 370842

- ○ Open Mon to Fri 9-5; Sat and Sun 9-4.
  Other times on request.
- £ Admission free; donations welcome.
- ☞ Situated behind the shops at Port Clarence, 3 miles from A19 on the A1046. Bus: No. 51 from Stockton.

Facilities: 🅿 Car and coach parking 🚻 Toilets 🍽 Refreshments 🛝 Play area.

Facilities for Disabled: ♿ Wheelchair access. Trail for the visually handicapped in preparation.

Restrictions: 🐕 Dogs on leads.

Description: The farm is at present a six-acre traditional livestock farm. There is a wide selection of rare breeds of livestock, most of which are bred on site. The farm's conservation area has just been acknowledged as the second best in the North-East. It contains a large pond for dipping, as well as flower meadows and hedgerows. Many activities are suitable for children including an adventure play area and a pets corner. You can try your hand at milking a goat, and watch working sheepdogs in action. The farm is an ideal educational day out, and provides classrooms which can each accommodate up to thirty children. The farm is staffed by workers and volunteers, and members of the local community are encouraged to become involved.

On-Site Activities: Guided walks; talks on and off the farm site.

Special Events: Barbecues; help-out days; family fun days.

Educational Facilities: Education officer/centre; school visits. Large classrooms. Telephone in advance for full information.

NORTHUMBERLAND
Chillingham Wild Cattle Park

Chillingham Wild Cattle Association, c/o Estate House, Chillingham, Alnwick, Northumberland NE66 5NW   Tel: (06685) 250

- ○ Open daily 10-12 and 2-5 Apr to end Oct. Pre-booked groups only on Tues and Sun mornings.
- £ Adults £2.00, children 50p. Group rates: £1.50.
- ☞ Take A1 to Alnwick then via Eglingham, or A697 to Wooler then via Chatton. Bus: Alnwick to Wooler bus.

Facilities: 🅿 Car and coach parking 🚻 Toilets 🛍 Gift/book shop.

Restrictions: 🐕 No dogs.

Description: This 360-acre park offers a unique opportunity to see the true descendants of the original ox which roamed these islands before the dawn of history. The park is a living museum, showing how Northumberland looked in bygone centuries. It is also the home of the Chillingham wild white cattle, known to be the sole remaining pure-bred herd of this species in the world. Access to the park is only available through continuous conducted tours led by the park warden who not only relates the history and habits of the herd but also a little of the estate's history and the make-up of the surrounding landscape. Other wildlife which can sometimes be seen includes fallow deer, foxes, badgers and red squirrels.

On-Site Activities: Guided tours for all visitors.

Educational Facilities: School visits.

Northumberland National Park

Northumberland National Park and Countryside Department,
Eastburn, South Park, Hexham, Northumberland NE46 1BS
Tel: (0434) 605555

Description: The Northumberland National Park and countryside comprises some of the wildest and most beautiful scenery in England, as well as some important historical sites such as Hadrian's Wall. The area covers the stunning heritage coastline, the lovely North Pennines and the wild Cheviot Hills and North Tyne Valley.

There are numerous organised activities to make your visit even more enjoyable: you can go pond dipping, do photographic and art workshops, the more adventurous can go canoeing or rock climbing. Organised walks cater for everyone from toddlers and the disabled to seasoned hikers. If you prefer to go it alone, the bird life is very diverse, and birdwatching is popular here. There are also numerous walks and trails for mountain bikes.

TYNE AND WEAR
Washington Waterfowl Park

District 15, Washington, Tyne and Wear NE38 8LE Tel: (091) 416 5454

- ○ Open daily 9.30-5 in summer, 9.30-4 in winter.
- £ Adults £2.95, children £1.50, family £7.50, OAPs £2.20. Group rates: on application.
- ☞ Follow signs off A195 and A1231 to centre 4 miles from A1M or 1 mile from A19. River Wear ferry from Sunderland. Bus: Nos. 186 from Washington, or X4 from Newcastle to Sunderland Mon to Sat.

Facilities: 🅿 Car and coach parking 🚻 Toilets ☕ Refreshments.

Facilities for Disabled: ♿ Wheelchair access; flat, good paths suitable for wheelchairs and pushchairs.

Restrictions: 🚫 No dogs.

Description: This 100-acre centre is home to well over 1,000 ducks, geese and swans. There is a series of specially planned walkways and secluded hides from which the visitor can admire some of the world's most endangered species of wildfowl. The reserve boasts the largest colony of grey herons in the country and the wild bird feeding station regularly attracts bullfinches, great spotted woodpeckers and sparrowhawks. A spectacular flock of Chilean flamingoes is as at home here as any of the trumpeter swans, king eiders, emperor geese and even the colourful oystercatchers. This is the place for the keenest of birdwatchers as well as for those just wanting a peaceful stroll. Children will be delighted by the many birds that will feed from their hands.

On-Site Activities: Talks and education programme by arrangement with the centre staff.

Educational Facilities: Education officer/centre; school visits. Schools membership scheme running with educational programmes in line with national curriculum. School's Out programme of activities for children in school holidays.

NORTH-WEST ENGLAND

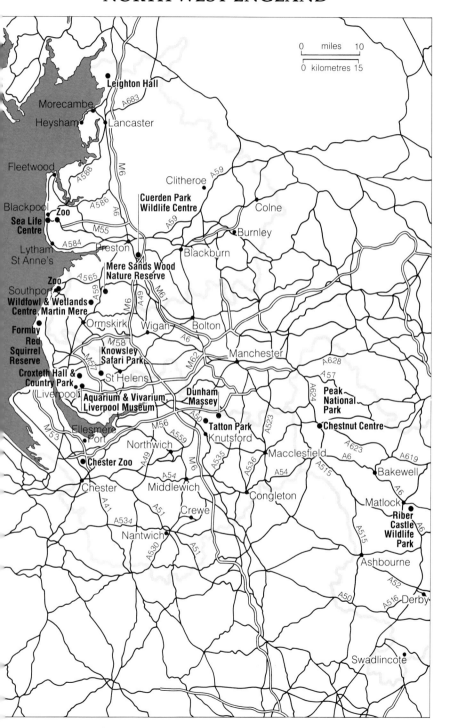

CHESHIRE
Chester Zoo

Upton-by-Chester, Cheshire CH2 1LH Tel: (0244) 380280

- ○ Open daily from 10; closing times vary seasonally. Closed Christmas Day.
- £ Adults £5.50, children and OAPs £3.00, family (2 + 3) £17.00. Group rates: adults £4.50, children £2.50.
- ☞ Off A41, 2 miles north of Chester city centre.
  Bus: No. C40 from Town Hall bus exchange.
  Rail: Bache station (20 minutes walk).

Facilities: 🅿 Car and coach parking 🚻 Toilets – mother and baby facilities ☕ Refreshments 🛍 Gift/book shop 🛝 Play area.

Facilities for Disabled: ♿ Wheelchair access. Trail for visually handicapped by arrangement.

Restrictions: 🚫 No dogs.

Description: Possibly the UK's largest garden zoo set in 110 acres, Chester Zoo has an extensive collection of animals and a formidable reputation for breeding. There are 5,000 animals, 500 species, with a special emphasis on conservation. Amongst the most endangered species to see are orang-utans, Waldrapp ibis, Rodrigues fruit bats, Colombian black spider monkeys and partula snails. All are on the red data book listing and have bred successfully since being at Chester Zoo. One of the most recent births was the baby black rhino in 1991, a species which is virtually extinct in the wild. The zoo also has the largest social group of chimpanzees in the UK. The visitor is able to see an entire family group from the grandmother to grandchild and listen to their chatter over loudspeakers in the chimp house. The large tropical house has one of Europe's leading collections of birds including one of the best breeding collections of parrots in the country.

On-Site Activities: Guided tours; brass rubbing; meet the keeper; monorail.

Educational Facilities: Education officer/centre; school visits.

Dunham Massey

Dunham Massey Hall, Altrincham, Cheshire WA14 4SJ Tel: (061) 941 1025

○ Garden open daily 12-5 Apr to Oct; house, shop and restaurant open Sat to Wed 12-5 Apr to Oct.

£ House and garden: adults £4.00, children £2.00, family £10.00. Group rates (over 15): £3.50. Garden only: adults £2.00, children £1.00. Admission free for National Trust members.

☞ M6 junction 19, M56 junction 7/8, off A56 3 miles south-west of Altrincham. Bus: North Western/GM No. 38. Rail: Altrincham station then bus to Dunham.

Facilities: Car and coach parking Toilets – disabled facilities Refreshments Gift/book shop.

Facilities for Disabled: Wheelchair access to garden, limited access to house. Toilets. Braille literature.

Restrictions: Dogs in park only and must be on leads. No photography in house.

Description: Fallow deer live in the grounds and visitors can walk all round the woodland and pasture areas. Over 100 different species of wild bird have been spotted here. Cormorants, little grebes and grey herons regularly visit the pond areas. Woodpeckers – the green, great spotted and lesser spotted varieties – are often seen in the woodlands, as well as tawny owls, little owls, sparrowhawks, grey partridges and pheasants. Five species of bat fly around the grounds and for those visitors with a keener eye there is a profusion of insects to see. Over eighty varieties of moth and more than 180 species of wood-dwelling beetle have all been recorded here. There are also many rare butterflies, dragonflies and damselflies.

On-Site Activities: Bat walks; wildlife walks; children's guidebook and quiz.

Special Events: Craft fairs, concerts etc.

Educational Facilities: Education officer/centre; school visits. Special school activities by prior arrangement. Educational visits can be pre-arranged in which the children participate in housekeeping and cooking as it was done in the past.

Tatton Park

Knutsford, Cheshire WA16 6QN Tel: (0565) 750250

- ○ Open daily 10.30-6 Apr to end Oct; Tues to Sat 11-5 Nov to Mar.
- £ Admission free to park. Car parking (to Apr 1993) £2.00. Explorer tickets (for all attractions): adults £6.00, children £4.20, family £18.50. Group rates: 20% discount.
- ☞ M6 junction 19, M56 junction 7 then follow signs. Rail: Knutsford station (3 miles).

Facilities: ℙ Car and coach parking ♦♦ Toilets ☕ Refreshments 🎁 Gift/book shop ⛰ Play area.

Facilities for Disabled: ♿ Limited wheelchair access. Trails for visually handicapped in planning stage.

Restrictions: 🐕 No dogs in gardens or mansion.

Description: England's most complete historic estate, Tatton Park offers five centuries of living history. The Old Hall was the focal point of Tatton village until the end of the seventeenth century, and serves as a vivid reminder of those times. The park forms the grounds of a neo-classical mansion house which was built in the last century. The house is open to the public, and its display of original furnishings paints a picture of how life was conducted by the aristocracy and their servants. There is also a farm where rare breeds of pigs, horses, poultry and cattle live together in a working environment. The park itself consists of 1,000 acres of country, where red and fallow deer roam free throughout the whole park. Two lakes are home to wildfowl; and a variety of other birds live in the grounds. Tatton has particularly good educational facilities, and is well equipped to host school parties. There is a programme of environmental activities in the park where children are invited to spend a day with the rangers exploring the animal kingdom.

On-Site Activities: More or less anything by prior arrangement.

Special Events: Calendar of events available on request.

Educational Facilities: Education officer/centre; school visits. Special programmes for schools.

DERBYSHIRE
Chestnut Centre

Castleton Road, Chapel-en-le-Frith, Derbyshire SK12 6PE Tel: (0298) 814099

- ○ Open daily 10.30-5.30 Mar to Dec; weekends only Jan and Feb.
- £ Adults £3.00, children £2.00.
- ☞ Follow signs from Chapel-en-le-Frith or Castleton.

Facilities: 🅿 Car and coach parking 🚻 Toilets ☕ Refreshments 🛍 Gift/book shop.

Restrictions: 🐕 Guide dogs only.

Description: For twenty-five years this centre has been concerned with the plight of the barn owl and the otter, both endangered species. There are several otter enclosures where visitors can see British otters, Canadian otters and Asian short-clawed otters playing or swimming in their pools. Many of the tawny owls and little owls at the centre have been rescued from the wild after injury. They are treated at the centre's hospital and then, when possible, released back into the wild. There is also the chance to see more exotic species like the Indian eagle owl and the great horned owl. An extensive nature trail meanders through the fifty-acre park and woodland. From here visitors can spot a variety of wild woodland birds and animals, sometimes even wild foxes and stoats.

On-Site Activities: Guided tours by arrangement.

Educational Facilities: Education officer/centre; school visits.

Peak National Park

Peak Park Joint Planning Board, Aldern House, Baslow Road, Bakewell, Derbyshire DE4 1AE   Tel: (0629) 814321

Description: The National Parks are a very special part of our heritage. The Peak National Park became the first National Park in 1951. It is surrounded by the industrial cities of the North and the Midlands. The Peak District was the last green unspoilt island at the southern end of the Pennines. It covers 555 square miles and takes in an area which straddles six counties. There are, in fact, two Peak Districts: the White Peak in the centre and south where the pearly grey limestone breaks the surface, and the Dark Peak where millstone grit forms the underlying rock of high moorlands. Amongst these varied landscapes there is an abundance of wildlife. On the moors, birds of prey can be found; in the bogs, frogs and toads; whilst on the peaks there are many species of bird, including the dipper. The botany of the area is quite extraordinary. Apart from the wild flowers of the peak, mosses, liverworts, lichens, toadstools and mushrooms can be found in the ancient woodland. Visitors may also go to the park to admire pretty villages, keeping alive ancient traditions.

Riber Castle Wildlife Park

Riber Castle, Matlock, Derbyshire DE4 5JU Tel: (0629) 582073

○ Open daily 10-5 in summer, 10-3.30 or 4.30 in winter. Closed Christmas Day.

£ Adults £3.50, children £2.00, OAPs £3.00. Group rates: adults £3.00, children £1.50.

☞ From M1 junction 28 follow A38 southbound then A615 to Matlock and towards Tansley. Turn right on to Alders Lane/Carr Lane for 1 mile. Bus: Matlock bus station. Walk across High Leys Park, up Church Street and follow public footpath on left between Highfields School and St Giles School; steep walk brings you out at entrance gate to park. Rail: Matlock station and walk (as above).

Facilities: 🅿 Car and coach parking 🚻 Toilets ☕ Refreshments 🎁 Gift/book shop 🎪 Play area.

Facilities for Disabled: ♿ Wheelchair access.

Restrictions: 🐕 No dogs round animal enclosures but welcome on car park, field and picnic areas.

Description: This wildlife park is home to many rare and endangered species. There are rare breeds of goat, sheep and deer as well as many more exotic animals like wallabies, racoons and Arctic foxes. Otters play and swim in large ponds and waterfalls. There is also the Rabbit Patch where a whole host of different breeds of rabbit can be seen. There is an impressive collection of birds of prey, everything from buzzards, falcons, hawks and kites to owls, eagles and even vultures. The biggest draw to the park must be the lynxes, one of the largest collections in the world. There are Siberian lynxes, caracal lynxes, European lynxes and bobcats. The park boasts a very successful breeding programme and several lynxes have been reintroduced into the wild.

On-Site Activities: Daily meet-a-keeper sessions (with owls, snakes, tarantulas or others). Formal talks for group visits by arrangement.

Educational Facilities: Education officer; school visits. Marble rubbing centre.

LANCASHIRE
Blackpool Zoological Gardens

East Park Drive, Blackpool, Lancashire FY3 8PP Tel: (0253) 65027

- ○ Open daily 10-5.15 (last admission) in summer, 10-4 in winter throughout the year except Christmas Day.
- £ Adults £3.50, children and OAPs £1.75, family £8.50. Group rates: adults £2.10, children £1.05.
- ☞ M6 junction 32 then follow M55 then A6 to Blackpool. Zoo signposted at end of Garstang Road. Bus: No. 21 from Tower, bus stop in Adelaide Street.

Facilities: 🅿 Car and coach parking 🚻 Toilets – mother and baby facilities ☕ Refreshments 🛍 Gift/book shop 🛝 Play area.

Facilities for Disabled: ♿ Wheelchair access.

Restrictions: 🐕 No dogs.

Description: Just a few minutes from the seafront of Britain's busiest resort, the thirty-two landscaped acres of Blackpool Zoo are home to over 400 species from all over the world. They include big favourites like the lion, tiger, giraffe, elephant and gorilla. Also popular are feeding times for the sealions and penguins which take place twice daily. Visitors may also ride on the zoo's miniature railway.

On-Site Activities: Sealion and penguin feeding times; miniature railway.

Educational Facilities: Education officer/centre; school visits. Educational talks.

North-West England – Lancashire   103

Cuerden Park Wildlife Centre

Lancashire Wildlife Trust, Cuerden Park, Shady Lane, Bamber Bridge, Preston PR5 6AU   Tel: (0772) 324129

○   Open daily throughout the year.
£   Admission free.
☞   From M6 junctions 28 and 29 or M61 junctions 8 and 9.

Facilities: ◨ Car parking ♦♦ Toilets – disabled facilities.

Facilities for Disabled: ♿ Wheelchair access; some paths have steps; kissing gates are designed to let wheelchairs through; access to the lake can be arranged by the warden for disabled fishermen. Toilets.

Restrictions: 🐕 Dogs under control.

Description: Set in 600 acres of the valley of the River Lostock, this rich and mixed habitat of woods, grassland, ponds and river contains a wide variety of wildlife and there is an extensive network of footpaths.

By the lake, great crested grebes, herons, coots, moorhens and mallards can often be seen, as well as kingfishers. Great spotted woodpeckers, long-tailed tits, tree creepers and goldcrests can be found in the woods, and there are nuthatches in the wood by the park centre.

Most of the ungrazed grassland is cut for hay, encouraging wild flowers and butterflies. In spring, orange-tipped butterflies are attracted by the lady's smock. Later arrivals are the meadow and wall brown, common blue, small copper and large skipper. In summer, there are also many damselflies and dragonflies.

On-Site Activities: Guided walks, events and conservation work parties organised by the warden.

Special Events: Wildlife festival in the autumn; WATCH club parties and picnics.

Educational Facilities: School visits by arrangement with the warden.

Leighton Hall

Carnforth, Lancashire LA5 9ST   Tel: (0524) 734474

- ○   Open Tues to Fri and Sun from 2 May to end Sep; open at 10 for school parties.
- £   Adults £3.00, children £1.90, OAPs £2.50. Group rates: £2.50, schools £1.50.
- ☞   M6 junction 35; follow signs from A6 junction with M6 north of Carnforth. Rail: Silverdale station (2 miles).

Facilities: ⓟ Car and coach parking 🚻 Toilets 🍴 Refreshments 🛍 Gift/book shop.

Facilities for Disabled: ♿ Wheelchair access.

Restrictions: 🐕 No dogs in garden, but welcome on leads in 80-acre park.

Description: Leighton Hall has the largest collection of trained birds of prey in the north of England. Different species of owl, kestrel, eagle and buzzard are flown regularly in the grounds of the hall. The two trained falconers give comprehensive talks to the public with a strong emphasis on conservation. Even when the birds are not flying, visitors can wander around the estate and see anything from a steppe eagle to a lanner falcon on display. Many of the native birds are brought to the hall after they have suffered injuries in the wild. The falconers nurse them back to health and then release them. A nature trail winds through the woodland where visitors can spot rabbits, ducks and roe deer before entering a walled garden where there is a newly constructed open maze.

On-Site Activities: Guided tours of hall. Displays with birds of prey plus commentary on open days.

Educational Facilities: Education officer/centre; school visits. Special educational programme during mornings.

Mere Sands Wood Nature Reserve

Holmeswood Road, Rufford, Lancashire L40 1TG Tel: (0704) 821809

○ Open daily 9-5 throughout the year except Christmas Day.

£ Admission free. Parking free for Lancashire Wildlife Trust members, £1.00 others.

☞ From A59 Rufford village, follow B5246 ½ mile towards Southport. Bus: Southport-Chorley and Burscough-Preston to Rufford. Rail: Preston-Ormskirk railway to Rufford station (¾ mile).

Facilities: 🅿 Car parking 🚻 Toilets.

Facilities for Disabled: ♿ Wheelchair access.

Restrictions: 🐕 Dogs on leads. Visitors must keep to paths. Bank holiday parking for members of Lancashire Wildlife Trust only.

Description: Mere Sands Wood covers over 100 acres of woodland, lakes and heaths and is a rich haven for wildlife. This is a Site of Special Scientific Interest owned and run by the Lancashire Wildlife Trust, and very important for wintering waterfowl including teal, goldeneye, pintail and shoveler, and breeding species such as great crested and little grebes. There are seven hides with plans for a new underwater hide and education wing. Visitors may see red squirrels, kingfishers, and a great range of dragonflies and butterflies.

On-Site Activities: Guided walks for groups are available by prior arrangement. Participation work days.

Special Events: Annual bat and hedgehog days; annual countryside weekend.

Educational Facilities: Education centre; school visits.

Sea Life Centre, Blackpool

The Promenade, Blackpool, Lancashire FY1 5AA Tel: (0253) 22445

- ○ Open daily from 10 throughout the year.
- £ Adults £3.95, children £2.95.
- ☞ Follow signs on promenade.

Facilities: Toilets Refreshments Gift/book shop Play area.

Facilities for Disabled: Wheelchair access.

Description: The Blackpool Sea Life Centre is situated on the promenade and has the largest collection of living sea creatures in the UK. With over 2,000 indigenous sea creatures on permanent display, the large open-topped displays give visitors a fascinating viewing angle of life beneath the waves. Apart from local fish such as cod, pollack, ling and conger eel, visitors can see a large number of tropical and reef fish including a Queensland grouper and three Australian cleaner wrasse. The main feature of the centre is the sharks, and they are contained in a 100,000-gallon tank. Blackpool has the largest tropical shark display in Europe with over thirty sharks of eighteen different species including horn, bull-headed, bamboo and epaulette. The epaulette shark is considered one of the most interesting as it uses its four paddle-like fins to move along the bottom of the tank. A 'scratch card' marine trail adds extra fun, as does the high-tech atmospheric sound interpretations.

On-Site Activities: Feeding displays; touch-pool talks; regular talks and demonstrations.

Educational Facilities: Education officer; school visits. Annual teachers' evenings. Special project packs available which are tailored to the national curriculum.

The Wildfowl & Wetlands Centre, Martin Mere

Martin Mere, Burscough, Ormskirk, Lancashire L40 0TA   Tel: (0704) 895181

○   Open daily 9.30-5.30 in summer, 9.30-4.30 in winter.

£   Adults £3.50, children £1.75, family £8.75, OAPs and students £2.50. Group rates: adults £2.80, children £1.30, OAPs £2.00.

☞   Signposted from M61, M58 and M6. Off A59 6 miles from Ormskirk and 10 miles from Southport.
Bus: No. 3 from Ormskirk (not Sun).

Facilities: 🅿 Car and coach parking 🚻 Toilets – disabled and mother and baby facilities ☕ Refreshments 🛍 Gift/book shop ⛹ Play area.

Facilities for Disabled: ♿ Wheelchair access. Braille trail. Toilets and changing room facilities.

Restrictions: 🐕 No dogs.

Description: A marshland reserve of 360 acres, recreated from derelict farmland in the early 1970s, Martin Mere supports 132 species of waterfowl, many of which will feed straight from your hand! Permanent home to 120 different kinds of wildfowl, Martin Mere is the regular wintering ground for thousands of ducks, geese and swans such as whooper swans from Iceland and Bewick swans from Siberia. Other species include great and Chilean flamingoes; mandarin ducks; laysan teals; goosanders; greylag, red-breasted and magpie geese; ruddy, eider, shoveler and Carolina ducks. A network of carefully planned walks, with secluded hides and a genuine Norwegian log cabin complete with turfed roof, affords views of countryside and many wild plants, butterflies, insects and birds.

On-Site Activities: Guided walks, children's activities including brass rubbing and badge-making. Play areas and hides, floodlit swans.

Special Events: Eco-fun morning. School holiday activities.

Educational Facilities: Education officer/centre; school visits. Educational programmes in the discovery centre.

MERSEYSIDE
Aquarium & Vivarium, Liverpool Museum

National Museums and Galleries on Merseyside, William Brown Street, Liverpool L3 8EL Tel: (051) 207 0001

- ○ Open Mon to Sat 10-5, Sun 12-5 throughout the year.
- £ Admission free.
- ☞ Situated in Liverpool city centre; well signposted.

Facilities: 🅿 Car and coach parking (limited; pay and display car park) 🚻 Toilets - disabled and mother and baby facilities ☕ Refreshments 🛍 Gift/book shop.

Facilities for Disabled: ♿ Wheelchair access. Toilets.

Restrictions: 🐕 Guide dogs only.

Description: The Aquarium and Vivarium forms part of the Liverpool Museum complex. Over 1,000 creatures from around the world, many of them rare and unusual, are displayed in twenty-six large tanks in the aquarium and ten in the vivarium. They include yellow surgeon fish, clown fish and long fire eels which are hand-fed every day. There is a Mexican red-leg tarantula and an octopus. They are kept in authentic reconstructions of different habitats, together with life-like replicas of animals and plants which are otherwise too difficult to keep in captivity. There are also displays of fish and invertebrates from around the British Isles.

Educational Facilities: Education officer/centre; school visits. Lecture rooms available by arrangement. Factsheets and other literature available.

Croxteth Hall & Country Park

Muirhead Avenue, Liverpool L12 0HB Tel: (051) 228 5311

○   Country park open daily 9-7 or dusk throughout the year. Farm open daily 11-5 throughout the year. Hall and walled garden open daily 11-5 Easter to end Sep.

£   Country park admission free. Farm, hall and walled garden: adults £2.00, children £1.00, concession £1.00. Group rates: 10% discount for pre-booked groups.

☞   Follow signs from junction of M57 and A580 (10 minutes) or from A5058 Liverpool ring road (Queens Drive). Bus: No. 61 to West Derby village (1 mile) from Liverpool city centre.

Facilities: P Car and coach parking ⫯ Toilets ⏣ Refreshments 🛍 Gift/book shop ⛰ Play area.

Facilities for Disabled: ♿ Limited wheelchair access. Trail for visually handicapped by arrangement.

Restrictions: 🐕 Guide dogs only in formal areas or indoors.

Description: The country park has 500 acres of parkland to explore including woodland with a variety of wildlife. The home farm has a unique collection of rare breeds: pigs, cattle, sheep, poultry and baby animals. There is a Victorian farmyard where visitors have the opportunity of contact with the animals. You can also see farming demonstrations.
    The historic hall has displays to show a country house at the turn of the century. A traditional Victorian walled garden and warm greenhouses edged by unusual fruit trees are also there to be explored. There is an adventure playground and miniature railway, and all ages can learn to ride at the Earl's former stables.

On-Site Activities: Full programme of events, nature trails and guided walks, most with a countryside or wildlife theme.

Special Events: Living history theme weeks throughout the year and special events in summer.

Educational Facilities: Education centre; school visits.

Formby Red Squirrel Reserve

Victoria Road, Freshfield, Formby, Merseyside L37 1LJ   Tel: (07048) 78591

○   Open daily during daylight hours throughout the year.

£   Per car £1.60 Apr to end Oct; £1.00 Nov to end Mar weekdays or £1.60 weekends. Coaches £10.00 all year.

☞   2 miles off A565, 15 miles north of Liverpool, 2 miles west of Formby. Bus: Merseyside Circular Nos. 4A, 161/4. Rail: Freshfield station (1 mile).

Facilities: Car and coach parking Toilets Gift/book shop.

Facilities for Disabled: Wheelchair access; hard surface paths to red squirrel reserve and Cornerstone walk.

Restrictions: Dogs on leads.

Description: Almost 500 acres of dunes and woodland have been bought by the National Trust and now form Formby Point. There is a wealth of native wildlife, much of it rare, such as the red squirrel and natterjack toad. On the beach, sand, mud and silt provide feeding grounds for gulls and wading birds like oystercatchers and sanderlings. The shifting dunes are stabilised by marram grass, and the troughs between the sandy ridges gather enough freshwater to sustain the natterjack toads. Further inland, the old dunes are much more stable and a succession of plants, including asparagus fields, build up into the woodland of birch, alder and pine.

Every so often, winter storms reveal the ancient footprints of creatures which roamed the coast thousands of years ago. Great elk and buffalo tracks have been seen, as well as those of a man wearing moccasins! A great place to see wildlife, but access is restricted to keep erosion to a minimum.

On-Site Activities: Guided walks.

Educational Facilities: School visits.

Knowsley Safari Park

Prescot, Merseyside L34 4AN   Tel: (051) 430 9009

- ○ Open daily 10-4 Mar to end Oct.
- £ Per car £8.00. Extra attractions paid for individually.
- ☞ M57 junction 2 then follow signs.
  Bus/Rail: to Prescot (10 minutes).

Facilities: 🅿 Car and coach parking 🚻 Toilets – mother and baby facilities ☕ Refreshments 🛍 Gift/book shop 🎠 Play area.

Facilities for Disabled: ♿ Wheelchair access.

Restrictions: 🐕 No dogs; kennels available.

Description: The park is based around a five-mile drive through many animal enclosures. There is a herd of African elephants, a breeding pride of lions and some tigers. There is a wooded area with deer, blackbuck and wallabies. A large mixed display area contains animals from all around the world: camels from Arabia, guanacos from South America, zebras and wildebeest from Africa and many more. The baboons will delight in trying to dismantle your car as you pass through the monkey jungle! Huge bison and white rhinos finish off the drive-through; look out for baby rhinos as they breed here. Other attractions include a pets corner where you will be allowed to touch the animals and a children's amusement park. There is also a miniature railway, reptile house and sealion shows.

Educational Facilities: School visits.

Southport Zoo and Conservation Trust

Princes Park, Southport, Merseyside PR8 1RX Tel: (0704) 538102

○ Open daily 10-6 in summer, 10-4 in winter.

£ Adults £2.40, children £1.40, OAPs £1.90.
Group rates (over 25): adults £2.00, children £1.20.

☞ Follow signs to Southport town centre.

Facilities: 🅿 Car and coach parking 🚻 Toilets
🍴 Refreshments 🎁 Gift/book shop 🛝 Play area.

Facilities for Disabled: ♿ Wheelchair access.

Restrictions: 🐕 Dogs on leads.

Description: Set amidst landscaped gardens, the zoo is home to over 800 animals. Cats include the lion, lynx, serval, leopard cat and the ocelot. A successful breeding programme has been set up for the chimpanzee and lar gibbon and the zoo can boast one of the world's largest breeding groups of mandrills in captivity. Other mammals include the South American tapir, the coati, otters, capybara and wallabies. A number of birds can be discovered: the flamingo, penguin, rhea. Other attractions include a reptile house which is home to a caiman and a number of snakes, an aquarium, parrot house, pets corner barn and an invertebrate collection including some bird-eating spiders.

On-Site Activities: Talks given by prior arrangement.

Educational Facilities: Education centre; school visits.

SOUTH-EAST ENGLAND

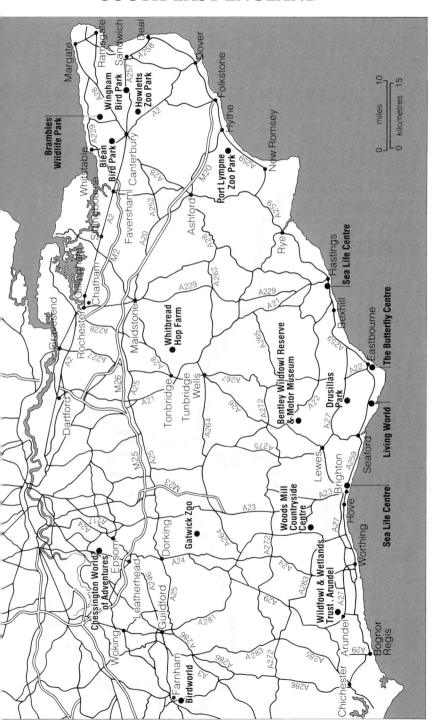

EAST SUSSEX
Bentley Wildfowl Reserve & Motor Museum

Halland, near Lewes, East Sussex BN8 5AF Tel: (0825) 840573

- ○ Open daily 10.30-4.30 early Mar to end Oct, weekends Nov and Dec, Feb and early Mar; closed Jan.
- £ Adults £3.10, children £2.40, OAPs £2.40, family (2 + up to 4) £8.00. Group rates and concessions: on application.
- ☞ Follow signs off A22, A26 and B2192 to 7 miles north-east of Lewes.

Facilities: 🅿 Car and coach parking ⚥ Toilets 🍽 Refreshments 📚 Gift/book shop 🛝 Play area.

Facilities for Disabled: ♿ Wheelchair access.

Restrictions: 🐕 Dogs in parking area only.

Description: The Bentley estate covers 100 acres of Sussex countryside. Attractions include a mansion, a motor museum and an extensive reserve with ponds, lakes and over 1,000 waterfowl. The estate features a woodland activity trail where children and adults alike can learn more about the forest. An adventure playground gives the younger ones a chance to enjoy themselves, and there is a small animals section where children can play with an assortment of farmyard friends. A miniature steam railway operates throughout the summer. The motor museum houses an impressive collection of Bentley classic cars as well as many other makes.

On-Site Activities: Nature trail. Guided tours and talks by arrangement.

Educational Facilities: Education officer/centre; school visits.

The Butterfly Centre

Royal Parade, Eastbourne, East Sussex BN22 8NH Tel: (0323) 645522

○ Open daily 10-5 throughout the year.
£ Adults £2.25, children £1.25, family £6.50. Group rates: on application.
☞ At eastern end of Eastbourne seafront. Bus: local service.

Facilities: 🅿 Car parking 🍵 Refreshments 🛍 Gift/book shop.

Facilities for Disabled: ♿ Wheelchair access.

Restrictions: 🐕 No dogs.

Description: A butterfly safari park! Visitors can stroll around tropical glasshouse gardens which provide free-flying butterflies from all over the world with a natural habitat. There are 500 exotic butterflies and over thirty different species. All stages of breeding, including egg-laying, caterpillar-rearing and courtship displays among the tropical flowers, can be closely observed and photographed. The centre also has separate educational and pictorial displays.

Educational Facilities: School visits.

Drusillas Park

Alfriston, East Sussex BN26 5QS Tel: (0323) 870234

○ Open daily 10.30-5 in summer, 10.30-4 in winter. Restaurant and shops open until 6.
£ Adults £4.50, children £3.95. Group rates: adults £3.25, children £2.95. Several concessions.
☞ On A27 between Eastbourne and Brighton. Bus: Nos. 125, 725, A2, 499 from Lewes or 725, 726, 499 from Eastbourne. Rail: Lewes, Berwick or Polegate stations (6, 1½ and 3½ miles).

Facilities: 🅿 Car and coach parking 🚻 Toilets – disabled and mother and baby facilities 🍵 Refreshments 🛍 Gift/book shop 🎠 Play area.

Facilities for Disabled: ♿ Wheelchair access to all areas; wheelchair hire. Toilets. Rail carriage for disabled visitors.

Restrictions: 🐕 Dogs only in gardens and grounds, not in zoo or adventure playground.

Description: Set in the beautiful South Downs of Sussex, Drusillas' emphasis is on providing a specialist zoo for children and families. Working beavers, playful otters, families of monkeys, farm animals and exotic animals can all be seen in careful recreations of their natural habitats. Barn owls live in a barn, tawny owls in a woodland setting and snowy owls amongst the pines, rocks and waterfalls. Emus and wallabies graze together while flamingo lagoon teems with wildfowl. Penguins can be observed underwater and animals such as meerkats are displayed in family groups. A variety of hands-on learning material adjoins each enclosure and quizzes and working models encourage children to assimilate information while having fun. A roaring dinosaur model moves its head and munches a fern at the edge of a pond of live crocodiles. An acre of the park is devoted to play equipment, many of which are 'real' items such as buses, jeeps, boats and fire engines. The play areas are divided up into all age ranges and there is an indoor play barn for wet weather activities. Other attractions include the rainforest story, a working pottery, rose gardens, a Japanese garden and locomotives that travel round the park through animal paddocks and a tunnel.

On-Site Activities: Train rides; adventure playground; quizzes and working models.

Educational Facilities: Education officer/centre; school visits.

Living World

Seven Sisters Country Park, Exceat, Seaford, East Sussex BN25 4AD
Tel: (0323) 870100

○ Open daily 10-5 mid-Mar to end Oct; weekends and holidays in winter.

£ Adults £2.20, children £1.30, family (2 + 2) £6.20, OAPs £1.30. Group rates: £1.30.

☞ On A259, 2 miles east of Seaford, 5 miles west of Eastbourne. Rail: Seaford station (2 miles). Bus: No. 712.

Facilities: 🅿 Car and coach parking ♦♦ Toilets 🍴 Refreshments 👍 Gift/book shop.

Facilities for Disabled: ♿ Wheelchair access.

Description: The Living World is set within the Sussex Downs area of Outstanding Natural Beauty. The zoo specialises in smaller animals but all are fascinating and unusual. They include butterflies, stick insects, scorpions, reptiles, spiders and ants. There are about forty living displays of both exotic and native animals. The Seven Sisters Park covers 700 acres of beautiful chalk downland with its rare plants and butterflies, cliffs and beach. The saltmarsh and lagoon attract abundant wildlife, especially sea and wetland birds, that visitors can enjoy. The cliffs provide excellent nest sites for jackdaws and fulmars. You can also go horse riding, canoeing, camping, walking, swimming and fishing.

On Site Activities: Guided walks and lectures.

Special Events: Mini-beast handling events have been very popular and attracted interest from national TV and press.

Educational Facilities: School visits are a major feature of the park.

Sea Life Centre, Brighton

Madeira Drive, Brighton, East Sussex BN2 1TB Tel: (0273) 604234

○ Open daily 10-6 throughout the year; last admissions at 5.

£ Adults £4.25, children £3.25, OAPs £3.75. Group rates: on application.

☞ Well signposted in Brighton.

Facilities: Toilets   Refreshments   Gift/book shop   Play area.

Facilities for Disabled: Wheelchair access.

Description: The Sea Life Centre aquarium in Brighton holds 130,000 gallons of sea water! There are dozens of British sharks to be seen as well as rays and conger eels. An unusual display is the replica of a deep sea submarine complete with deep sea diver in a heavy duty diving suit, working on an underwater pipeline. The diver's oxygen supply going into his suit is also a cleverly designed means of oxygenating the surrounding water. As with all the Sea Life Centres, there are many imaginative marine life displays to be seen and much to learn about not only the inland waters of Britain but also creatures of the deep sea.

On-Site Activities: Feeding displays; touch-pool talks; regular talks and demonstrations.

Educational Facilities: Education officer/centre; school visits. Annual teachers' evenings; special project packs available which are tailored to the national curriculum.

Sea Life Centre, Hastings

Rock-a-nore Road, Hastings, East Sussex TN34 3DW Tel: (0424) 718776

○ Open daily 10-6 throughout the year; last admissions at 5.

£ Adults £3.95, children (4-14) and students £2.75, OAPs £3.45. Group rates (10 or over): on application.

☞ Well signposted in Hastings.

Facilities: 🅿 Car and coach parking 🚻 Toilets
  ☕ Refreshments 🛍 Gift/book shop ⛟ Play area.

Facilities for Disabled: ♿ Wheelchair access.

Description: A popular attraction here is a particularly large touch-pool facility where children and adults alike can get a hands-on introduction to the creatures that occur along the British shoreline, including hermit crabs, starfish, shore crabs, velvet swimming crabs and spiny spider crabs. One weird and wonderful creature to look out for is the sea mouse, an iridescent marine worm which resembles a rainbow-coloured, furry snail! A real treat for visitors is the diving demonstration: a diver goes into the water to feed the fish, which include sharks and trigger fish, and sometimes some of the diners will even take food from the mouth of the diver.

On-Site Activities: Feeding displays; touch-pool talks; regular talks and demonstrations.

Educational Facilities: Education officer/centre; school visits. Annual teachers' evenings. Special project packs available which are tailored to the national curriculum.

KENT
Blean Bird Park

Honey Hill, Blean, Canterbury, Kent Tel: (0227) 471666

○ Open daily 10-6 in summer, 10-5 in winter.

£ Adults £2.75, children £1.75, concession £2.50. Group rates: 20% discount.

☞ Halfway between Canterbury and Whitstable on A290. Bus: No. 605 from Canterbury bus station.

Facilities: 🅿 Car and coach parking 🚻 Toilets – disabled and mother and baby room 🍽 Refreshments 🛍 Gift/book shop 🛝 Play area.

Facilities for Disabled: ♿ Wheelchair access. Toilets.

Restrictions: 🐕 No dogs.

Description: Blean claims to have one of the largest collections of parrots on show in the country. There are also owls, pheasants, peacocks, herons and hornbills to be seen, and a children's zoo with a host of small animals. These include alpacas, coatimundis, rabbits and wallabies, along with traditional farm animals such as pigs, cows, goats and chickens.

On-Site Activities: Nature trail; parrot hand-rearing demonstration and talk; children's zoo.

Special Events: Falconry display some weekends; Easter egg hunts; teddy bears' picnic.

Educational Facilities: Education officer/centre (summer 1993); school visits.

South-East England – Kent

Brambles Wildlife Park

Herne Common, Herne Bay, Kent CT6 7LQ Tel: (0227) 712379

○ Open daily 10-5 Easter to end Oct.

£ Adults £2.20, children £1.20, OAPs £1.50.
Group rates (over 20): adults £1.20, children £1.00.

☞ On A291 between Herne Bay and Canterbury.
Bus: Nos. 604, 606, 607 from Herne Bay or Canterbury.

Facilities: 🅿 Car and coach parking ♯ Toilets – mother and baby facilities ☕ Refreshments 🛍 Gift/book shop 🛝 Play area.

Facilities for Disabled: ♿ Wheelchair access.

Restrictions: 🐕 Guide dogs only.

Description: Set in twenty acres of natural woodland, you can spot much of the native wildlife you would expect to find such as squirrels, green and great spotted woodpeckers, hedgehogs etc. You are guaranteed to see the following native and foreign wildlife in captivity: badgers, fallow deer, buzzards, kestrels, Scottish wildcats, red foxes and barn owls. Exotics include sika deer, silver foxes, black swans, racoons, wallabies, maras and guanacos.

Educational Facilities: School visits.

Howletts Zoo Park

Bekesbourne, near Canterbury, Kent CT4 5EL Tel: (0227) 721286

- ○ Open daily 10-5 in summer, 10-3.30 in winter (last admissions).
- £ Adults £6.50, children (4-14) £4.50. Group rates (over 20): adults £5.00, children £3.00.
- ☞ Off A2 Dover road 3 miles south-east of Canterbury follow signs. Rail: Bekesbourne station (15 minutes walk).

Facilities: Car and coach parking Toilets – mother and baby facilities Refreshments Gift/book shop Play area.

Facilities for Disabled: Wheelchair access.

Restrictions: No dogs. Children under 15 must be accompanied by an adult.

Description: John Aspinall's well-known wildlife park provides a chance to view animals from all over the world in spacious natural enclosures, where the animals are free to roam in and out of their shelters - and hide from visitors if they wish. Its colony of gorillas is the world's largest breeding group in captivity, and they can often be seen playing with their keepers. The zoo has African elephants; its clouded leopards are breeding successfully; and there are Indian and Siberian tigers, capuchins, lemurs, plus the world's first honey badgers, banded langurs and more besides. The aim is to breed such rare and endangered species and to return many to the wild.

On-Site Activities: Most weekends the Aspinall family can be seen playing with or feeding tigers, elephants, gorillas etc. Times are not specific.

Special Events: Summer puppet shows with a conservational theme. New Zoom Club opened for children.

Educational Facilities: Education officer/centre (opening 1993); school visits.

Port Lympne Zoo Park, Mansion and Gardens

Lympne, near Hythe, Kent CT21 4PD Tel: (0303) 264647

- ○ Open daily 10-5 in summer 10-3.30 in winter (last admissions).
- £ Adults £6.50, children (4-14) £4.50. Group rates (over 20): adults £5.00, children £3.00.
- ☞ Off A20 between Sellindge and Hythe. M20 junction 11. Bus: No. 10 from Ashford to Folkestone.

Facilities: Car and coach parking Toilets – mother and baby facilities Refreshments Gift/book shop Play area.

Facilities for Disabled: Wheelchair access; ask at entrance.

Restrictions: No dogs. Children under 15 must be accompanied by an adult.

Description: John Aspinall's second wildlife park is based on the same idea as Howletts. Rare breeds enjoy spacious enclosures: on view are the world's second largest group of Przewalski's horses and Barbary lions (both extinct in the wild). Along the two and a half mile zoo trek, visitors can see buffaloes roam, along with bison, deer and the West's only pair of Sumatran rhinos. There are more than fifty species including Indian and Siberian tigers, wolves, monkeys and snow leopards, and a major attraction is the gorilla pavilion. The 300-acre site is housed around the recently restored Sassoon Mansion, and its terraced gardens overlook the Romney Marsh and English Channel.

On-Site Activities: Safari trailer in season for pre-booked groups.

Special Events: Summer puppet shows with a conservational theme. Tours on the mansion in summer. New Zoom Club opened for children.

Educational Facilities: Education officer/centre; school visits.

Whitbread Hop Farm

Beltring, Paddock Wood, Kent TN12 6PY Tel: (0622) 872068

○   Open daily 10-6 in summer, 10-4 in winter throughout the year except Christmas Day and Boxing Day.

£   Adults £2.00, children, OAPs and disabled £1.50. Group rates: on application.

☞   From M20 junction 2 follow signs to Paddock Wood; farm is situated on B2015.

Facilities: 🅿 Car and coach parking 🚻 Toilets – disabled and mother and baby facilities ☕ Refreshments 👍 Gift/book shop 🛝 Play area.

Facilities for Disabled: ♿ Wheelchair access. Toilets.

Restrictions: 🐕 Dogs on leads

Description: The Whitbread Hop Farm offers a unique opportunity to explore the history of British hop growing and picking. It also boasts the largest collection of Victorian oasthouses in the world. Once a major industry in the region, it now hosts the hop story exhibition which depicts the lifestyle of people in bygone times. Shire horse activities can be seen at the famous Whitbread Shire Horse Centre. Birds of prey, such as snowy owls, European owls, eagle owls and kestrels are on view and owl flying displays are given. Other attractions include the animal village where touching and feeding animals such as sheep, goats and donkeys is encouraged. There is also a nature trail featuring many of the area's native flora and fauna.

Educational Facilities: School visits.

Wingham Bird Park

Busham Road, Wingham, Canterbury, Kent CT3 1JL Tel: (0227) 720836

○ Open daily 10-6 Mar to Oct; weekends 10-5 Nov to Feb.

£ Adults £2.50, children £1.75, disabled visitors half price. Group rates: 20% discount.

☞ On A257 Canterbury to Sandwich road.
Bus: Nos. 613 or 614 to Sandwich.

Facilities: 🅿 Car and coach parking 🚻 Toilets 🍽 Refreshments 🛍 Gift/book shop 🛝 Play area.

Facilities for Disabled: ♿ Wheelchair access.

Restrictions: 🐕 No dogs.

Description: The park boasts a varied collection of native and exotic birds: waterfowl, cockatoos, macaws, owls, soft-billed birds, lorys, lorikeets and many more, all housed in spacious flights. Wingham also has a large walk-through aviary of parrots, with soft-bills and smaller birds in the partitioned orchard aviary. A great deal of emphasis is placed on captive breeding projects for endangered bird populations. A tree-lined lake provides the setting for both waterfowl and picnics. Recent additions to the park include an area with rare breeds of farm animals, a pets corner and a children's playground.

Educational Facilities: Education officer; school visits.

SURREY
Birdworld

Holt Pound, Farnham, Surrey GU10 4LD Tel: (0420) 22140

- ○ Open daily 9.30-6 in summer, 9.30-3.30 in winter.
- £ Adults £3.65, children £2.00, family £10.70.
- ☞ On A325 road 3 miles south of Farnham.
  Rail/Bus: Aldershot station then bus to Birdworld.

Facilities: 🅿 Car and coach parking 🚻 Toilets – mother and baby facilities ☕ Refreshments 🛍 Gift/book shop 🛝 Play area.

Facilities for Disabled: ♿ Wheelchair access.

Restrictions: 🐕 No dogs.

Description: An outstanding collection of over 250 species of bird; special emphasis is placed here on breeding rare species. Everything from pheasants and ducks to cranes, rheas and ostriches, not to mention a whole host of rare parrots, are hatched and reared in the incubator research station. There is a comprehensive array of birds of prey – owls, hawks, eagles, falcons and vultures – as well as many more surprising species like ground hornbills, secretary birds and the black and white casqued hornbills. Penguin Island is home to a colony of Humboldt penguins which produce over twenty youngsters a year. But the most popular aviary at Birdworld is probably the Seashore Walk where the visitor can see shags, gannets and pelicans swimming underwater. The Tropical Walk gives opportunities to see some of the more exotic species, and visitors can spot many of our native birds from the Woodland Trail.

Special Events: Baby bird days – children can see young birds at close quarters and have the chance to speak to their keepers.

Educational Facilities: Education officer; school visits. Educational question books and topic packs available.

South-East England – Surrey 127

Chessington World of Adventures
Leatherhead Road, Chessington, Surrey KT9 2NE Tel: (0372) 727227

○ Open daily 10-5.30; last admissions at 3.

£ Adults £10.75, children (4-14) £9.75, OAPs £5.00.
  Group rates (20 and over): on application.

☞ On A243, follow signs from M25 junction 9.
  Rail: Chessington South station (10 minutes walk).

Facilities: 🅿 Car and coach parking 🚻 Toilets – disabled and mother and baby facilities ☕ Refreshments 👍 Gift/book shop 🛝 Play area.

Facilities for Disabled: ♿ Wheelchair access. Toilets.

Restrictions: 🐕 No dogs or other pets; dogs must not be left in cars.

Description: In this large theme park-cum-zoo is Chessington Zoo – not a large one but still an interesting collection of some of the best-known and loved zoo animals. There are penguins, giraffes, lions, tigers, jaguars, snow leopards and camels. The ape house contains gorillas and chimpanzees and, in a long enclosure at the other end of the zoo, are the lar gibbons. In the monkey walk are Celebes macaques, capuchins, Barbary apes and others. The polar bear enclosure allows for underwater viewing and it is possible to catch them taking a dip. There is a large and varied collection of birds from familiar species like flamingoes, parrots and storks to more unusual ones like ibis and Andean condors. A new reptile house is entitled and decorated as Snakebite Saloon and houses several large snakes including the yellow anaconda. Above the zoo is a monorail safari skyway with a recorded commentary on the animals beneath.

Educational Facilities: School visits.

Gatwick Zoo

Russ Hill, Charlwood, Surrey RH6 0EG Tel: (0293) 862312

- ○ Open daily 10.30-6 Mar to end Oct; weekends 10.30-4 Nov to Feb.
- £ Adults £3.50, children £2.50, OAPs £3.00 except Sun and bank holidays. Group rates (over 20): 10% discount.
- ☞ A217 or A23 north of Gatwick airport then follow signs to Charlwood and zoo.

Facilities: 🅿 Car and coach parking 🚻 Toilets – disabled facilities ☕ Refreshments 👍 Gift/book shop 🛝 Play area.

Facilities for Disabled: ♿ Wheelchair access. Toilets.

Description: The zoo is set amidst ten acres of gardens and is home to a number of small mammals and birds. A monkey island surrounded by a moat holds spider and squirrel monkeys (other monkeys to be found include marmosets and lemurs). Muntjac deer are also featured. A variety of birds are held in aviaries (cockatoos, rare poultry, soft-billed birds) and macaws fly free through the zoo grounds. There are a number of birds of prey including snowy and eagle owls and there is a penguin enclosure. The tropical house holds a variety of tropical plants and birds. There is also a butterfly house. Other features include an education centre, picnic area, play area for children and a gift shop.

On-Site Activities: Free guide books; many walk-through enclosures. Free talks to groups of over thirty visitors.

Educational Facilities: Education officer/centre; school visits.

WEST SUSSEX
The Wildfowl & Wetlands Trust, Arundel

Mill Road, Arundel, West Sussex BN18 9PB Tel: (0903) 883355

○ Open daily 9.30-6.30 in summer, 9.30-5 in winter.

£ Adults £3.75, children £1.90, family £9.40, OAPs, £2.80. Group rates: adults £3.00, children £1.50, OAPs £2.25; school parties £1.30.

☞ Follow signs from A27 Arundel. Bus: open-top bus service from Brighton during the summer. Rail: Arundel station (1½ miles).

Facilities: P Car and coach parking Toilets Refreshments Gift/book shop.

Facilities for Disabled: ♿ Wheelchair access.

Restrictions: Guide dogs only.

Description: Over 1,000 birds find a home at this beautiful wetland site. Either out of doors or from the large viewing gallery, visitors can spot a whole array of ducks, geese and swans. At about four o'clock, many birds, like Bewick swans, smews, tufted ducks and pochards, all eagerly await the feed barrow containing a menu of bread, wheat, fish and lettuce! Such rich pickings encourage the birds to display some very poor table manners as they squabble over tasty morsels! The viewing gallery often has displays of environmental topics as well as craft activities for children, puzzles and exhibits. Guided walks can be arranged, followed by a relaxing meal or snack in the Eider restaurant or coffee shop. The shelduck shop offers gifts as well as a host of natural history books, binoculars and telescopes to let you view the birds in close-up.

On-Site Activities: Guided walks and talks can be arranged for groups if pre-booked.

Special Events: Events calendar available from the centre if you send a stamped addressed envelope; events organised through the year, particularly in the school holidays.

Educational Facilities: Education officer/centre; school visits.

Woods Mill Countryside Centre

Woods Mill, Henfield, West Sussex BN5 9SD Tel: (0273) 492630

○ Open Sat 2-6, Sun and bank holidays 11-6 Easter to end Sep; Tues, Wed, Thurs, Sat 2-6 and Sun and bank holidays 11-6 in school holidays.

£ Adults £2.00, children £1.00, family £5.00, OAPs £1.50. Group rates: 10% discount if booked in advance.

☞ On A2037 Henfield to Shoreham road 1½ miles south of Henfield. Bus: from Horsham and Brighton to Henfield (1½ miles). Rail: Brighton (10 miles), Hassocks (8 miles).

Facilities: 🅿 Car and coach parking 🚻 Toilets ☕ Refreshments 🛍 Gift/book shop.

Facilities for Disabled: ♿ Wheelchair access to reserve only.

Restrictions: 🐕 Guide dogs only.

Description: The centre is housed in an eighteenth-century water mill surrounded by a fifteen-acre nature reserve. A nature trail winds through the reserve, passing woodland, meadow, marsh, reedbeds and a lake. Twenty species of dragonfly have been recorded here, together with sightings of kingfisher, great spotted woodpecker and the white admiral butterfly. Printed trail guides are available for both adults and children and nets are provided at the pond-dipping area.

Inside the mill there is a countryside exhibition, a vivarium with harvest mice, an observation beehive (weather permitting) and an eight-metre replica of an oak tree. The mill's water wheel and machinery have been restored and audio-visual presentations can be seen.

On-Site Activities: Guided walks at weekends or by appointment. Self-guided nature trail.

Special Events: WATCH day in September; wildlife explorers in August.

Educational Facilities: Education officer/centre; school visits for booked parties only. Day courses for schools throughout the year. Custom-built classroom is available for school parties and day courses.

SOUTHERN

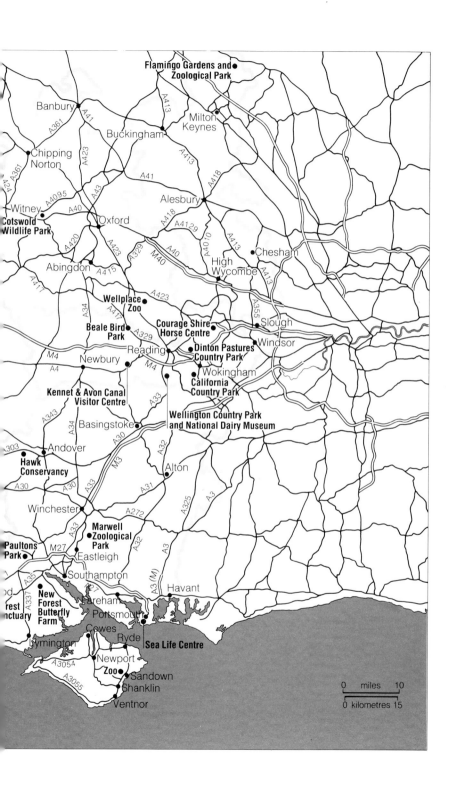

BERKSHIRE
Beale Bird Park

Church Farm, Lower Basildon, Reading, Berkshire RG8 9NH
Tel: (0734) 845172

- ○ Open daily 10-6 Mar to end Sep, 10-5 in winter with limited facilities.
- £ Adults £4.00, children (3-16) and disabled visitors £2.00, OAPs £3.00. Group rates: on application.
- ☞ Follow signs from M4 junction 12. Bus: No. 5 Reading to Oxford. Rail: Pangbourne station (1 mile).

Facilities: P Car and coach parking ♿ Toilets – disabled and mother and baby facilities ☕ Refreshments 🛍 Gift/book shop ⛱ Play area.

Facilities for Disabled: ♿ Wheelchair access. Toilets.

Restrictions: 🚫 No dogs.

Description: The park is set amidst the Chilterns and is home to a number of exotic animals. The bird collection can boast more than 150 species, ninety per cent of which are captive-bred and reared. The tropical house holds fish, reptiles, insects and marmosets. Other animals to be found include llamas and Vietnamese pot-bellied pigs. Statues are spread through the gardens and a collection of model ships and boats is displayed within the pavilion. There are numerous walks, picnic areas and an adventure playground. If you're feeling lazy, it is possible to take a ride on the narrow gauge railway. Fishing is also available.

Educational Facilities: School visits.

Dinton Pastures Country Park

Davis Street, Hurst, Reading, Berkshire RG10 0TH Tel: (0734) 342016

○ Open daily throughout the year.

£ Admission free.

☞ From M4 junction 10, follow signs to Woodley. Park is off B3030 Twyford Road ½ mile from Winnersh crossroads and BR station.
Rail: Winnersh station (½ mile).

Facilities: 🅿 Car and coach parking 🚻 Toilets – disabled facilities ☕ Refreshments 🎁 Gift/book shop 🛝 Play area.

Facilities for Disabled: ♿ Wheelchair access to most of the park including Sandford Bird Hide. Toilets. Reduced rates for disabled anglers; good access for disabled visitors to fish.

Restrictions: 🐕 Dogs under control.

Description: The 230 acres of Dinton Pastures in the Loddon Valley are on the site of an old dairy farm. The area has inherited a rich variety of habitats while more recent gravel extraction has left eight lakes which make up more than half the area of the park. The resulting landscape is a mosaic of lakes, rivers, meadows and hedgerows, ideal for quiet country walks and for spotting wildlife along the way. You can see swans, geese, ducks, dragonflies, squirrels, bats and butterflies. There is a visitor centre with displays and information and plenty of special events thoughout the year for adults and children.

Special Events: Bird and wildlife walks; orienteering; teddy bears' picnic; night walks; Easter craft fair; rambles; lakeside concerts. Countryside ranger events throughout the year for seven to fourteen-year-olds. Countryside workshops can be hired for parties or groups.

Educational Facilities: Education officer/centre; school visits.

California Country Park

Nine Mile Ride, Finchampstead, Wokingham RG11 4HT Tel: (0734) 730028

- ○ Open daily throughout the year.
- £ Admission free.
- ☞ Park is situated between Wokingham and Sandhurst. From A321, B3016 or A3095 join Nine Mile Ride and follow signs for Finchampstead. Follow park signs from California Crossroads double roundabout.

Facilities: 🅿 Car and coach parking 🚻 Toilets – disabled and mother and baby facilities ☕ Refreshments 👍 Gift/book shop 🎠 Play area.

Facilities for Disabled: ♿ Wheelchair access to most of park. Toilets.

Restrictions: 🐕 Dogs under control.

Description: The park provides 100 acres of peaceful countryside and a valuable habitat for wildlife. The lake is home to ducks, geese and swans while other small woodland animals such as woodpeckers, squirrels and deer live among the trees. You can walk on special paths through Longmoor Nature Reserve which includes an ancient bog which is a Site of Special Scientific Interest.

Special Events: Activity days; guided bog walks; bat barbecue; teddy bears' picnic; night walks.

Educational Facilities: Education officer/centre; school visits.

Courage Shire Horse Centre

Cherry Garden Lane, Maidenhead Thicket, Maidenhead, Berkshire
Tel: (0628) 824848

○   Open daily 10.30-5 Mar to end Oct; last admissions at 4.

£   Adults £2.50, children and OAPs £2.00. Group rates (over 10): adults £2.50, children and OAPs £1.85.

☞   M4 junction 8/9 then follow A4 towards Reading. Bus: Reading Bee Line or Reading to London Goldline X1 Express.

Facilities: P Car and coach parking Toilets – mother and baby facilities Refreshments Gift/book shop Play area.

Facilities for Disabled: Wheelchair access. Dray rides for disabled where possible.

Restrictions: Dogs on leads.

Description: Visitors to the Courage Shire Horse Centre can wander around or take a guided tour to meet these magnificent and dignified horses as well as learn and experience a little of their history. They are show horses and a team of experts keep them in peak condition. Just beyond the stables is a working forge where a farrier can be watched, normally three days a week, keeping the Shires shod. He hammers red hot iron into shoes to fit each particular horse. Alternatively, a cooper or harness maker can be seen at work most days. There are dray rides and a small animal and bird area.

On-Site Activities: Guided tours.

Educational Facilities: School visits.

Kennet & Avon Canal Visitor Centre

British Waterways, Lower Wharf, Padworth, near Reading, Berkshire RG7 4JS
Tel: (0734) 712868

- Open Mon to Sat 10-5 except bank holidays.
- £ Admission free. School parties: on application.
- ☞ M4 junction 12 then A4 and A340 to Basingstoke. Rail: Aldermaston station (2 minutes walk).

Facilities: Car and coach parking Toilets Limited refreshments Gift/book shop Play area.

Facilities for Disabled: Wheelchair access.

Restrictions: No dogs.

Description: A canalman's cottage on the Kennet and Avon canal has been converted into an attractive canal-side visitor and information centre to provide details of the canal systems and boat hire. The one-mile self-guided tour along the canal bank and through neighbouring woodland may offer a glimpse of the native wildlife. The canal and banks are host to a wide variety of aquatic animals and plants. Approximately twenty species of dragonfly and damselfly frequent the area. Some of the most common are the banded demoiselle, common blue, blue-tailed, brown hawker and migrant hawker. The rich water vegetation hides an abundance of fish – pike, dace, carp, tench and rudd. Many species of bird can be seen and there are numerous mammals including vole, water shrew, otter and mink on the water's edge and wood mice, rabbits and bank voles among the hedgerows. The area is popular with the local walkers and day trippers and, if lucky, it is possible to see kingfishers, roe deer, nightingales, tufted ducks, mallards, herons and much more wildlife.

On-Site Activities: Self-guided trails; guided walks by arrangement; information display on canal and restoration.

Educational Facilities: School visits.

Wellington Country Park and National Dairy Museum

Riseley, Reading, Berkshire RG7 1SP Tel: (0734) 326444

○ Open daily 10-5.30 Mar to end Oct, weekends 10-dusk in winter.

£ Adults £2.75, children £1.25. Winter rates: adults £1.40, children 70p. Group rates: on application.

☞ M4 junction 11 then A33 towards Basingstoke or M3 junction 6 then A33 towards Reading. Follow signs to park on B3349.

Facilities: 🅿 Car and coach parking 🚻 Toilets – disabled and mother and baby facilities ☕ Refreshments 👍 Gift/book shop 🛝 Play area.

Facilities for Disabled: ♿ Wheelchair access. Toilets.

Description: The lakeside setting of the park extends to 600 acres of woodland, lake and parkland where you can see a range of waterfowl including ducks, swans, Canada geese and herons. Red and fallow deer roam in the deer park, while magpies, squirrels, woodpeckers and other birds find their home among the trees. In the small animals farm there are guinea pigs, rabbits, goats, sheep, pigs and calves, some of which can be stroked. The National Dairy Museum traces the development of the dairy industry in the UK. There is also a miniature steam railway, adventure playground and refreshment facilities.

On-Site Activities: Guided walks by arrangement.

Educational Facilities: Education officer/centre; school visits.

BUCKINGHAMSHIRE
Flamingo Gardens & Zoological Park

Western Underwood, Olney, Buckinghamshire MK46 5JR Tel: (0234) 711451

○   Open daily Easter to end Sep.

£   Adults £3.50, children £1.50, OAPs £2.50.
Group rates (over 10): adults £2.50, children £1.25.

☞   Off M1 at junction 14 to Newport Pagnell. Turn right beforeNewport Pagnell to Olney. Follow signs to Flamingo Gardens and Zoological Park.
Rail: Milton Keynes station (8 miles) then taxi.

Facilities: Car and coach parking Toilets Refreshments Gift/book shop Play area.

Facilities for Disabled: Wheelchair access.

Restrictions: No dogs.

Description: Situated in the peaceful village of Weston Underwood in pleasant natural surroundings, these bird gardens offer visitors the chance to see many rare birds from all round the world. There are endangered species of bird close to extinction that have been successfully bred within the park and in some cases, such as with the American bald eagle, it is the only place in the UK to have achieved captive breeding. The park has a wide variety of birds and there are over 150 species including toucans, parakeets, cockatoos, flamingoes, cranes, storks, eagles, vultures, pelicans and waterfowl with many tropical birds among these. Animals within the park comprise such impressive creatures as the American bison and red buffalo, antelopes, camels, llamas, white wallabies, maras and wild sheep. The Flamingo Gardens and Zoological Park is the home of the Pelican Trust and seven out of the eight species of pelican worldwide are to be found within the park's bird collection.

Educational Facilities: School visits.

DORSET
Abbotsbury Swannery

New Barn Road, Abbotsbury, near Weymouth, Dorset DT3 4JT
Tel: (0305) 871684

- Open daily 9.30-5 Easter to end Oct; by arrangement in winter.
- £ Adults £2.90, children £1.00, concession £2.60.
- On B3157 between Weymouth and Bridport. Rail/Bus: Weymouth and Dorchester stations; limited bus service to Abbotsbury.

Facilities: Car and coach parking Toilets Refreshments Gift/book shop.

Facilities for Disabled: Wheelchair access.

Restrictions: No dogs.

Description: The swannery is unique in that it contains the only managed colony of mute swans in the world. Visitors can mix freely with these magnificent birds. Walks through the reedbeds lead to hides overlooking the fleet lagoon. The oldest duck decoy still working in the UK dates from the seventeenth century. During the winter guided wildfowl tours can be arranged.

Twenty acres of natural woodland form the sub-tropical gardens, famous for magnolias, camellias, rhododendrons and many more plants benefitting from the mild position close to the sea. The tithe barn now houses the country museum which brings the old countryside alive; find out what it was like to be a shepherd, poacher, farrier or animal doctor in years gone by. The thatched village of inns and craft shops similarly harks back to the past.

On-Site Activities: Audio-visual centre; children's Ugly Duckling activity trail; guided walks.

Educational Facilities: Education officer; school visits.

Brownsea Island

The National Trust, Brownsea Island, Poole, Dorset BH15 1EE
Tel: (0202) 707744

- ○ Open daily 10-5 Apr to mid-Oct.
- £ Adults £2.00, children £1.00. Group rates (over 15 people) adults £1.80, children 90p.
- ☞ Boats from Poole or Sandbanks every 30 minutes.

Facilities: Toilets Refreshments Gift/book shop.

Facilities for Disabled: ♿ Wheelchair access. Braille literature.

Restrictions: No dogs.

Description: Covering over 500 acres of mainly heath and woodland, this island is owned by the National Trust. Here red squirrels can still be found as well as sika deer and peacocks. There are many species of bird to be seen, particularly in the 248 acres of reserve run by the Dorset Trust for Nature Conservation. The reserve is well wooded and contains a marsh, salt water lagoon and two lakes, supporting an interesting array of wildlife. A sanctuary for ducks, geese and waders, the reserve also has a flourishing heronry, a black-headed gullery and colonies of common and sandwich terns. A wonderful sight to greet visitors in spring is the masses of daffodils that cover large areas. Boats run frequently across to the island from Poole Quay and Sandbanks. Please note that no dogs are permitted on the island.

On-Site Activities: Guided tour of nature reserve.

Educational Facilities: Education officer/centre; school visits. Teacher pack available.

Dorset Heavy Horse Centre

Edmondsham Road, near Verwood, Dorset Tel: (0202) 824040

○ Open daily 10-5.30 Easter to end Oct; last admissions at 4.

£ Adults £2.95, children £1.95, family £8.00, OAPs £2.50. Group rates (children, schools and playgroups): £1.45

☞ Follow signs from crossroads at centre of Verwood (1 mile), 4 miles from Ringwood.

Facilities: 🅿 Car and coach parking 🚻 Toilets
 🍽 Refreshments 🎁 Gift/book shop 🅰 Play area.

Facilities for Disabled: ♿ Wheelchair access.

Restrictions: 🐕 Dogs on leads.

Description: The Dorset Heavy Horse Centre specialises in six different breeds of gentle giants: the Shire, Clydesdale, Suffolk Punch, Percheron, French Ardennes and Canadian-Belgian. In contrast, visitors can meet some of the smallest ponies in the world: there is a miniature Shetland stud complete with their own tiny stable yard. Displays of farm equipment add to the feel of a working farm. Visitors get a chance to feel real horsepower by ploughing and having wagon rides. Children's play areas together with pets and bird aviaries will add to youngsters' fun. There is also a café and picnic area, together with a souvenir shop.

On-Site Activities: Commentaries three times a day on the different breeds of heavy horses and miniature Shetland ponies. Wagon rides (weather permitting).

Special Events: Foals born each summer (times unpredictable!).

Educational Facilities: School visits. Video room can be used for school parties on wet days. Worksheets available for school parties on request.

The Natural World

Poole Aquarium Leisure Centre, The Quay, Poole, Dorset BH15 1JH
Tel: (0202) 686712

- ○ Open daily 9.30-9 in summer holidays, 10-5 in winter.
- £ Adults £2.95, children £2.00, family (2 + 2) £8.50. Group rates: adults £2.00.
- ☞ From Poole town centre, follow signs to Poole Quay. Bus: Arndale Centre bus station (15 minutes walk) or Poole Quay. Rail: Poole station (15 minutes walk) or bus to Poole Quay.

Facilities: 🅿 Public car and coach parking 🚻 Toilets ☕ Refreshments 🛍 Gift/book shop.

Facilities for Disabled: ♿ No wheelchair access. Physical help can be given to visitors with lightweight wheelchairs.

Restrictions: 🐕 Dogs on leads.

Description: Overlooking Poole harbour, the indoor aquarium and serpentarium displays some 160 different species of fish including three lemon and one nurse shark. You can attend shark feeding on Wednesdays, Fridays and Sundays. A large number of amphibians and reptiles are on show: tree frog, alligator, crocodile, plus an extensive collection of venomous snakes. Down in the snake pit, it is possible to handle a python and be photographed to prove it! A number of spiders, including a tarantula, are on display. Lectures are available for school parties. Other attractions include a model railway and restaurant.

On-Site Activities: Curator's talk lasting thirty minutes for groups of up to thirty-five at additional charge of £10.00. Shark feeding; model railway.

Educational Facilities: Education officer/centre; school visits. Exhibition to be extended to include a new classroom facility and an area demonstrating the effects of pollution and territorial destruction.

Radipole Lake RSPB Nature Reserve

Swannery Car Park, Weymouth, Dorset DT4 7TZ Tel: (0305) 778313

○ Open daily 8-dusk throughout the year.
 Shop open daily 9-5.

£ Part of the reserve requires permits for non-RSPB members: adults £1.50, children 50p, concession £1.00.

☞ Follow A354 or A353 to centre of Weymouth.
 Bus: Weymouth coach station (200 metres).
 Rail: Weymouth station (200 metres).

Facilities: 🅿 Car and coach parking 🚻 Toilets
 ☕ Light refreshments 🛍 Gift/book shop.

Facilities for Disabled: ♿ Wheelchair access.
 Trails for visually handicapped.

Restrictions: 🐕 Dogs restricted on part of reserve.

Description: This nature centre has a large viewing window with interpretative displays which looks over some of the 200 acres of reedbeds, lakes, pasture and scrub of the reserve. There are three hides, but two of them are restricted to RSPB members or by special permit. There are nature trails and guided walks are available. Birds which are regularly seen include mute swan, mallard, cormorant, great crested grebe and heron – all attracted by the water. Lucky visitors may see some rare birds like cetti's warblers, bearded tits, ring-billed and little gulls. In the autumn there are large numbers of reed warblers, and roosts of swallows and yellow wagtails. Interesting wetland plants also abound, such as sedge, reed mace and yellow iris. The habitat is carefully managed for the benefit of wildlife and plants. Rare species such as the marsh orchid and painted lady butterfly benefit from this work. Like most RSPB reserves, there is not only plenty of birdlife to see, but a whole lot more besides.

On-Site Activities: Weekly guided walks on Sunday afternoons. Guided walks for children.

Special Events: Details available from the nature centre.

Educational Facilities: Education officer/centre; school visits.

Sea Life Centre, Weymouth

17 Cobham Road, Ferndown, Dorset BH21 7PE Tel: (0202) 896289

○ Open daily from 10 throughout the year.

£ Adults £3.95; children £2.95. Group rates: on application.

☞ Follow signs.

Facilities: 🅿 Car and coach parking 🚻 Toilets
  ☕ Refreshments 🎁 Gift/book shop 🎠 Play area.

Facilities for Disabled: ♿ Wheelchair access.

Description: As well as offering a stunning variety of displays of British marine life, Weymouth has a very successful fish breeding unit with a number of success stories of young being reared and returned to the sea, including lumpsuckers and cuttlefish. Others that have been bred include the stickleback and pipefish and the centre carries out research into the breeding biology of many British marine creatures. There is also a unique quarantine facility which houses fish destined for other Sea Life Centres for a period during treatment for injury or disease; a sort of sea life hospital! The added attraction of Weymouth has to be its stunning living rainforest ecosystem filled with exotic tropical birds. There are even regular monsoons, three or four times daily!

On-Site Activities: Feeding displays; touch-pool talks; regular talks and demonstrations.

Educational Facilities: Education officer/centre; school visits. Annual teachers' evenings. Special project packs available which are tailored to the national curriculum.

Worldwide Butterflies and Lullingstone Silk Farm

Compton House, near Sherborne, Dorset Tel: (0935) 74608

○ Open daily 10-5 Apr to end Oct.

£ Adults £3.99, children £2.30, OAPs £2.99 or double ticket £5.00, family (2 + 2) £8.99. Group rates: on application.

☞ On A30 mid-way between Sherborne and Yeovil. Bus: from Sherborne or Yeovil.

Facilities: 🅿 Car and coach parking 🚻 Toilets – mother and baby facilities ☕ Refreshments 🛍 Gift/book shop 🛝 Play area.

Facilities for Disabled: ♿ Wheelchair access.

Restrictions: 🐕 Dogs only in car park.

Description: The butterfly house and silk farm are set within the grounds of Compton House, parts of which date back to the fourteenth century. Butterflies and huge moths flutter through a jungle of exotic plants. At most times of the year you can see eggs, caterpillars, chrysalids and adult butterflies all under one roof. Large beetles and camouflaged stick insects are less easy to spot than the colourful butterflies, but are worth looking out for. Outdoors there are butterfly gardens designed to attract native butterflies, as well as being attractive.

 The Lullingstone Silk Farm provided silk for the wedding dresses of Queen Elizabeth and the Princess of Wales. The farm grows mulberry trees, the food plant of the silk moth, and breeds the moths throughout the spring and summer. Visitors can see the silk produced being reeled (unwound from the cocoons). Giant silk moths like the huge giant atlas moth can also be seen in the rearing room.

Educational Facilities: School visits.

HAMPSHIRE
The Hawk Conservancy

Weyhill, Andover, Hampshire SP11 8DY Tel: (0264) 772252

- ○ Open daily 10.30-4 Mar to end Oct, open till 5 in summer.
- £ Adults £3.50, children (3-15) £1.75, OAPs £2.75. Group rates (20 or over): on application.
- ☞ Just off A303 about 3½ miles west of Andover. Rail/Bus: to Andover then local bus or taxi.

Facilities: 🅿 Car and coach parking ⚐ Toilets 🍴 Refreshments 🎁 Gift/book shop.

Facilities for Disabled: ♿ Wheelchair access.

Restrictions: ✘ No animals. Children under 15 must be accompanied by an adult.

Description: Of all the many species on view – falcons, owls, vultures, eagles, kites – by far the most popular are the owls which range in size from the spectacularly large European eagle owl to the diminutive scops owl. This is the most comprehensive collection of owls open to the public in the UK and many of the species are becoming increasingly rare in the wild. The centre believes that conservation is of paramount importance and aims to contribute by teaching people about these wonderful birds and how they live. Breeding birds are housed in special aviaries, and the centre operates a breeding scheme in co-operation with other zoos for species such as the milky eagle owl, spectacled owl and Falkland Islands caracara. As knowledge increases, they are constantly changing and improving their aviaries to meet the requirements of the birds. Flying demonstrations are given several times a day (weather permitting) so that visitors have an opportunity to appreciate the beauty and splendour of the birds in flight. Visitors may also have a chance to hold some birds and to take photographs. There is also an attractive heronry which attracts many species of wildfowl such as trumpeter swans and mandarin ducks.

On-Site Activities: Regular flying demonstrations (weather permitting) with opportunities to hold some of the birds at the end of the demonstration; information centre.

Educational Facilities: Education centre; school visits. Parties should book in advance to take advantage of educational tours.

Marwell Zoological Park

Colden Common, Winchester, Hampshire SO21 1JH Tel: (0962) 777406

- ○ Open daily 10-6 in summer, 10-5 in winter.
- £ Adults £5.50, children £4.00, OAPs £5.00. Group rates: £4.50.
- ☞ Take M3/M27. Situated 6 miles south of Winchester on B2177 Bishop's Waltham road. Rail: Winchester, Eastleigh and Hedge End stations (all about 6 miles).

Facilities: ▣ Car and coach parking ♦♦ Toilets – mother and baby facilities ⬛ Refreshments ♣ Gift/book shop ▲ Play area.

Facilities for Disabled: ♿ Wheelchair access. Trails for visually handicapped.

Restrictions: ✘ No dogs or pets.

Description: This is a world-famous wildlife park and there are around a thousand animals to see. Among the 130 species in the collection, many of which are endangered, are big cats, red pandas, zebras, antelopes, monkeys, hippos and rhinos. Young animals can be seen all the year round. New features are World of Lemurs and Encounter Village where visitors can make friends with rare and unusual domesticated breeds.

On-Site Activities: Animal handling; touch tables; behind the scenes tours; senior citizen specials; tours for the visually impaired.

Special Events: Winter wonderland event during December has to be pre-booked.

Educational Facilities: Education officer/centre; school visits.

New Forest Butterfly Farm

Longdown, Ashurst, near Southampton, Hampshire SO4 4UH
Tel: (0703) 292166

- ○ Open daily 10-5 Easter to end Oct.
- £ Adults £3.30, children £2.30, family (2 + 2) £9.70. Group rates: adults £3.00, children £2.00.
- ☞ Turn off the A35 between Southampton and Lyndhurst. Rail/Bus: Ashurst station then taxi (2 miles).

Facilities: 🅿 Car and coach parking ♀♂ Toilets – mother and baby facilities ☕ Refreshments 🛍 Gift/book shop 🛝 Play area.

Facilities for Disabled: ♿ Wheelchair access.

Restrictions: 🐕 No dogs.

Description: Set in the heart of the New Forest, the farm houses a huge indoor butterfly jungle where butterflies and moths live and breed in temperatures of up to 27°C. One section is devoted to tropical butterflies and another to their British cousins. Visitors in bright clothing may be mistaken for flowers! In the insectarium you can discover the delights of tarantulas, scorpions, praying mantids and stick insects – all behind glass, of course! The farm is surrounded by beautiful woodland and lays on rides in traditional horse-drawn farm wagons. More energetic visitors may prefer to explore the local countryside on foot. Other attractions include dragonfly ponds, aviaries and a children's adventure playground.

On-Site Activities: All groups given an introductory talk. Woodland walks.

Educational Facilities: Education officer; school visits.

New Forest Owl Sanctuary

Crow Lane, Crow, Ringwood, Hampshire BH24 1EA   Tel: (0425) 476487

○   Open daily 10-5 in summer, 10-4 in winter.

£   Adults £2.00, children £1.00, OAPs £1.50.
Group rates: on application.

☞   Follow signs off A31 Bournemouth to Southampton road about ½ mile from main road. Bus: from Ringwood every hour or taxi costs about £2.00.

Facilities: ▣ Car and coach parking ♦♦ Toilets – disabled and mother and baby facilities ⚌ Refreshments 👍 Gift/book shop ⚠ Play area.

Facilities for Disabled: ♿ Wheelchair access; ramps. Toilets.

Restrictions: 🐕 Guide dogs only.

Description: All those who love birds will be interested in this haven for their most notorious representatives. Set in the heart of the New Forest, the sanctuary houses many different species of owl and birds of prey. The staff have a keen interest in falconry and regular flying displays demonstrate the aerobatic skills and ferocity of some of the world's greatest hunting animals. You can compare the stoop of a falcon to the silent swoop of a barn owl. The sanctuary also features an education centre where children and adults can learn more about the life and habits of birds of prey and owls.

On-Site Activities: Talks; flying demonstrations both inside and out (weather permitting).

Educational Facilities: Education officer/centre; school visits. Bird of prey educational centre catered for about 96,000 people in 1991.

Paultons Park

Ower, near Romsey, Hampshire SO51 6AL   Tel: (0703) 813025

- ○ Open daily 10-5 Mar to Oct.
- £ Adults £5.75, children (under 14) £4.75.
  Group rates: on application, special rates for schools.
- ☞ M27 junction 2 A31/A36. Bus: Mon to Fri Wiltshire and Dorset buses, telephone (0722) 336855 for information. Sun in season Solent Blueline, telephone (0703) 226235 for information.

Facilities: 🅿 Car and coach parking 🚻 Toilets – mother and baby room ☕ Refreshments 🛍 Gift/book shop 🛝 Play area.

Facilities for Disabled: ♿ Wheelchair access.

Restrictions: 🐕 Guide dogs only.

Description: Paultons combines displays of wild creatures with a range of family leisure attractions. There is plenty to see, as over 100 species live on ponds, in the gardens or on the horseshoe-shaped lake. Its extensive collections of exotic birds and wildfowl include parrots, hornbills, emus, flamingoes, pelicans and owls. Pets corner has a variety of young animals including chipmunks, lambs, piglets and Belgian hares. The Rio Grande train takes visitors on a ride round the animal paddocks.

On-Site Activities: Land of the dinosaurs; magic forest; six-lane astroglide; bumper boats; go-karts; kid's kingdom with three acres of climbing activities.

Educational Facilities: School visits. Free education pack related to the national curriculum is available. Park also has a village life museum and Romany museum, a working Victorian watermill, Capability Brown gardens and a Japanese garden.

Sea Life Centre, Portsmouth

Clarence Esplanade, Southsea, Portsmouth, Hampshire PO5 3PB
Tel: (0705) 734461

○ Open daily 10-9 in summer, 10-6 in winter; last admission one hour earlier.

£ Adults £3.95, children £2.95. Group rates: on application.

☞ From M27 join M275 to Southsea then follow signs. Bus/Rail: Portsmouth and Southsea or Portsmouth Harbour stations then bus to Southsea seafront.

Facilities: 🅿 Car and coach parking 🚻 Toilets 🍽 Refreshments 🛍 Gift/book shop 🛝 Play area.

Facilities for Disabled: ♿ Wheelchair access.

Restrictions: 🐕 Guide dogs only.

Description: An extremely popular feature of Portsmouth Sea Life Centre is the Danger in the Depths exhibition. A dramatic living display includes some of the world's most deadly marine creatures such as the moray eel, the poisonous lion fish, electric rays and, of course, sharks! There is also a game or challenge for visitors: to identify correctly the different weapons used by the various examples of fish on display. Would you be a survivor of Danger in the Depths? Onlookers learn of the alternative devices used by an array of creatures for defence and predation. At the end of it all you can look forward to obtaining your trophy: the much-sought-after survivor's badge announcing 'I survived the dangers of the deep'!

On-Site Activities: Daily talks and feeding demonstrations.

Special Events: Birthday parties are available by pre-booking.

Educational Facilities: Education officer; school visits. All schools receive an individual talk designed for their needs.

ISLE OF WIGHT
Isle of Wight Zoo

Yaverland Seafront, Sandown, Isle of Wight PO36 8QB
Tel: (0983) 403883/405562

- ○ Open daily 10-6 Easter to Oct, Sun 10-4 in winter.
- £ Adults £3.99, children £2.99. Group rates: on request.
- ☞ On B3329. Bus: Nos. 8 and 44.

Facilities: 🅿 Car and coach parking 🚻 Toilets – mother and baby facilities ☕ Refreshments 👍 Gift/book shop 🛝 Play area.

Facilities for Disabled: ♿ Wheelchair access.

Restrictions: 🐕 No dogs.

Description: The two main specialisations of the zoo are the venomous snake centre and the big cat collection. At the snake centre, a large collection of particularly deadly snakes is held for the World Health Organisation and the zoo regularly supplies it with venom for medical research. The zoo is proud of its big cats but particularly its breeding pair of Bengal tigers which produced four cubs in the autumn of 1992. What is so remarkable, however, is the pair's track record of twenty-six cubs! It is looking increasingly likely that this is a world record for captive breeding and indeed the Guinness Book of Records is investigating. Other big cats include leopards, pumas and the smaller jungle cats. It is a small zoo which concentrates on giving a more personal touch to proceedings. Staff 'presenters' introduce visitors directly to the animals when possible. For example, there are snake and spider handling sessions, and at specific times of the year there is a chance to meet the baby animals. The zoo is on and around the historical sea front Granite Forest and adjacent to the 200-foot Culver Cliff which is rich in fossils.

On-Site Activities: Continuous programme of zoo chats and educational tours.

Special Events: Venomous snake handling and milking.

Educational Facilities: Education officer/centre; school visits.

OXFORDSHIRE
Cotswold Wildlife Park
Burford, Oxfordshire OX18 4JW Tel: (0993) 823006

○ Open daily 10-6 or dusk throughout the year.

£ Adults £4.00, children and OAPs £2.50.
Group rates: adults £3.20, children and OAPs £1.90.

☞ On A361 2 miles south of Burford. From M4 junction 15 or M5 junction 11 or M40 junction 8.

Facilities: 🅿 Car and coach parking 🚻 Toilets – mother and baby facilities ☕ Refreshments 🛍 Gift/book shop 🛝 Play area.

Facilities for Disabled: ♿ Wheelchair access.

Description: First opened in 1970, the Cotswold Wildlife Park is set in 120 acres of gardens and parkland and is dominated by a Gothic-style manor house, parts of which are open to the public. The park is home to a large and varied collection of animals from all over the world. The animal enclosures have been carefully designed and sited and many are particularly spacious; full advantage has been taken of the beautiful rural setting of the Cotswold Hills. Mammals include meerkats, otters, coatis, ruffed lemurs, monkeys and apes. Big carnivores are represented by two species of cat, the Persian leopard and Bengal tiger; and other big animals include white rhinos, Bactrian camels, scimitar-horned oryxes and zebras.

The tropical house exhibits a fascinating array of tropical plants, and living among the foliage are various insect, fruit and nectar-feeding birds, including tanagers from the Americas and white eyes from Africa and Asia. In the reptile house you will find iguanas, snakes and the giant Aldabran tortoise as well as tree frogs, fire salamanders and many species of fish. The gardens are well established and contain a large variety of named trees and shrubs together with extensive lawns which may be used for picnics.

Younger visitors are well catered for, not only in the children's farm, where some of the animals can be petted and stroked, but there is also an adventure playground and narrow gauge railway.

On-Site Activities: Talks can be booked in advance; children's farm; adventure playground; railway.

Special Events: Car rallies; bird of prey demonstrations; Morris dancing; archery tournaments.

Educational Facilities: Education officer/centre; school visits.

Wellplace Zoo

Ipsden, Oxfordshire OX9 6AD Tel: (0491) 680473

- ○ Open daily 10-5 Easter to end Sep; Sat and Sun 10-4 Oct to Easter.
- £ Adults £1.50, children 50p.
  Group rates (20 or over): adults £1.25, children 40p.
- ☞ Follow signs from A4074 or A423.

Facilities: 🅿 Car and coach parking 🚻 Toilets – disabled facilities ☕ Refreshments 🛍 Gift/book shop 🛝 Play area.

Facilities for Disabled: ♿ Wheelchair access. Toilets.

Restrictions: 🐕 No dogs.

Description: The zoo has a large and varied collection of animals from around the world. Many birds, such as eagle owls, penguins and parrots, breed each year. Visitors can feed the goats, sheep, rabbits, pheasants, guinea pigs, ponies and donkeys. More exotic creatures include otters, monkeys, llamas, serval cats, flamingoes, cranes, toucans and wallabies. There are even dinosaurs – well, large model ones at least! There are picnic areas, or you can eat at the coffee shop. As well as having fun in the play area, children can enjoy donkey rides.

On-Site Activities: Donkey rides.

Educational Facilities: School visits.

WEST COUNTRY

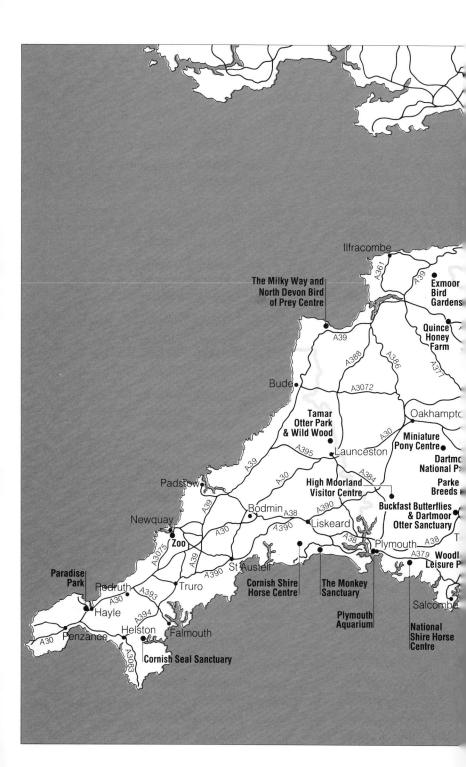

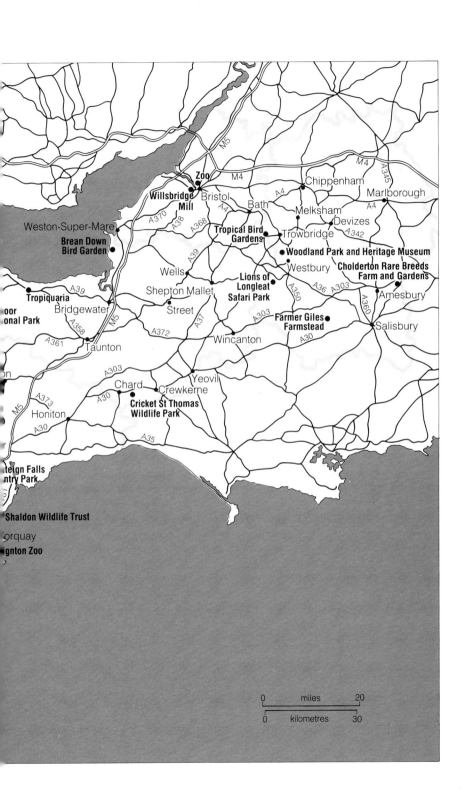

AVON
Bristol Zoo Gardens

Guthrie Road, Clifton, Bristol, Avon BS8 3HA   Tel: (0272) 738951

- ○ Open daily 9-6 in summer, 9-5 in winter. Closed Christmas Day.
- £ Adults £4.80, children £2.50, OAPs £2.50 on Mon only except bank holidays. Group rates: booking required, please telephone for information. Annual membership available.
- ☞ On A4176. M4 travellers join M5 and exit junction 17 southbound or junction 19 northbound and follow signs. Bus: Nos. 8, 9, 508, 509 (Templemeads), 582, 583 from Cheltenham Road, Broadmead, Kingswood.

Facilities: 🅿 Car and limited coach parking 🚻 Toilets – mother and baby facilities ☕ Refreshments 🛍 Gift/book shop 🛝 Play area.

Facilities for Disabled: ♿ Wheelchair access.

Restrictions: No pets.

Description: The twelve acres of gardens within Bristol Zoo are themselves an outstanding feature. Whatever the season there are colourful flowerbed displays, rock gardens and an exotic plant house to explore. An appealing exhibit at the zoo is the new lake island home for gibbons, marmosets, tamarins and lemurs, with unusual views of these animals afforded by a wooden walkway. The zoo's reptile house and aquarium offer the visitor stunning displays of animals from all over the world in very attractive, carefully laid-out settings. The tropical bird house is also full of luxuriant tropical plants and free-flying colourful birds can be viewed at close range. The zoo enjoys an international reputation for its captive breeding work including Persian leopards and Sumatran tigers. Field conservation projects are also supported both in this country and abroad, often with zoo staff having some direct involvement in the work. Why not try your hand at the new zoolympics trail? An opportunity to test your sporting skills against those incredible abilities within the animal kingdom.

On-Site Activities: Some feeding-time talks, touch tables, face painting, boat rides, bouncy giraffe.

Special Events: Easter festival: Easter bunny's egg race and events. Spring and summer festivals include Punch and Judy shows, face painting, boat rides, bouncy giraffe, keeper talks and animal encounters. Christmas festival includes Santa, craft fair, outdoor live nativity.

Educational Facilities: Education officer/centre; school visits. Conference facilities.

Tropical Bird Gardens

Rode, Bath, Avon BA3 6QW Tel: (0373) 830326

○ Open daily 10-6.30 or dusk throughout the year.
£ Adults £3.50, children (3-16) £1.75, OAPs £3. Group rates: on application.
☞ On A36 between Bath and Warminster.

Facilities: 🅿 Car and coach parking 🚻 Toilets – disabled and mother and baby facilities ☕ Refreshments 🛍 Gift/book shop ⛰ Play area.

Facilities for Disabled: ♿ Wheelchair access. Toilets.

Restrictions: 🐕 No dogs.

Description: The gardens at Rode are situated in seventeen acres of lovely natural surroundings. There are hundreds of brilliantly coloured exotic birds of about 230 different species, many very rare. The gardens were the first in Britain to breed scarlet ibis and umbrella cockatoos. Ornamental pheasants roam the gardens, macaws fly free amongst the trees, flamingoes, black-foot penguins, cranes and waterfowl are but a few of the other residents. Other attractions include woodland, flower gardens, a clematis collection, an ornamental lake, pets corner and a woodland steam railway. An information centre is on site.

Educational Facilities: School visits.

Willsbridge Mill

Avon Wildlife Trust, Willsbridge Mill, Willsbridge Hill, Bristol, Avon BS15 6EX Tel: (0272) 326885

- ○ Open Tues to Fri and Sun Easter to end Oct. Open bank holiday Mon.
- £ Adults £1.50, children £1.00, family (2 + 2) £4.00, concession £1.25. Group rates (15 or over): 25p off ticket price.
- ☞ Follow A431 from Bristol and Bath (6 miles) or A4175 from Keynsham. Signposted off A431 in Longwell Green. Bus: Badgerline No. 332 Bristol to Bath. Cityline No. 45 from Bristol. Rail: Keynsham station (1 mile).

Facilities: 🅿 Car parking 🚻 Toilets ☕ Refreshments 🛍 Gift/book shop.

Facilities for Disabled: ♿ Wheelchair access everywhere except part of nature reserve. Physical access improvements in nature reserve currently under way, to be followed by interpretative provision, in particular a 'cassette' trail.

Restrictions: 🐕 Dogs welcome.

Description: Within a twenty-two-acre nature reserve there are many varied wildlife habitats: woodland, meadow, quarry, scrub, streams and ponds, all supporting their own special and unique wildlife. The jigsaw trail is a self-guided nature trail and provides a circular route through the valley. Although there are no rare species, the main purpose of the reserve is to provide an opportunity for the visitor to get close to the everyday delights of nature. It could be an encounter with squirrels, bats, insects, kingfishers, dippers or amphibians. The mill has a range of free leaflets to encourage the visitor to explore. The boardwalk provides a chance to examine the underwater wildlife of the Syston brook. There are purpose-built teaching ponds for pond dipping. Another feature is the exciting hands-on wildlife and historical exhibitions in the impressively restored nineteenth-century corn mill which includes an observation beehive, stream life aquarium, spooky bat booth as well as plenty of buttons to press and wheels to turn.

On-Site Activities: Full events programme throughout season such as guided evening bat walks, guided dawn chorus walks, children's holiday activities, wildlife gardening demonstrations, supervised stream dipping, self-guided seasonal trails and environmental drama and art.

Educational Facilities: Education officer; school visits from March to November. An environmental education programme is provided and fulfils the requirements of the national curriculum in terms of environmental education.

CORNWALL
Cornish Seal Sanctuary

Gweek, near Helston, Cornwall TR12 6UG   Tel: (0326) 22361

- Open daily 9.30-6 Easter to Nov, 9.30-4.30 in winter.
- Adults £4.00, children £2.00, OAPs £3.00.
  Group rates: schools – adults £2.00, children £1.60; coaches on application.
- Follow A394 to Helston then A3083 towards the Lizard, then B3291 into Gweek.
  Rail: Truro, Redruth, Falmouth stations
      (14 miles, 11 miles, 8 miles).

Facilities: Car and coach parking   Toilets – mother and baby facilities   Refreshments   Gift/book shop   Play area.

Facilities for Disabled: Wheelchair access.

Restrictions: Dogs welcome.

Description: At the sanctuary, which is set in forty-five acres of landscaped grounds on the banks of the picturesque Helford River, there are ten pools and a well-equipped hospital for injured and abandoned baby grey Atlantic seals. As well as the grey seals, it also provides a home for other animals: some southern sealions, a Californian sealion, six Macaroni penguins and some abandoned donkeys. The centre's main aim is to rescue, care for, and return young seals to the wild – its success rate is about ninety per cent – but through its work it also offers a rare opportunity to see the beauty and gentle nature of these delightful creatures at close quarters. Pups can usually be seen between October and February. There is an exhibition centre with audio-visual displays about seals, dolphins and whales and the need for conservation of marine mammals.

On-Site Activities: Feeding times at 11 and 4 daily for about forty minutes are informative and entertaining.

Educational Facilities: School visits. Talks or lectures arranged. School packs available.

Cornish Shire Horse Centre

Trelow Farm, Wadebridge, Cornwall PL27 7RA Tel: (0841) 540276

○ Open daily 10-5 Easter to end Oct.

£ Adults £3.95, children £2.95.
Group rates: £1.00 per head discount.

☞ Off A39 Wadebridge to St Column Major road.
Bus: Western National.

Facilities: 🅿 Car and coach parking 🚻 Toilets – mother and baby facilities 🍴 Refreshments 🛍 Gift/book shop 🛝 Play area.

Facilities for Disabled: ♿ Wheelchair access.
Trails for the visually handicapped.

Description: There are over thirty Shire horses here: mares, foals and stallions. The indoor arena provides an all-weather means of meeting these gentle giants. Other animals include Shetland ponies, and at the small animals enclosure children can meet ducks, rabbits, dogs and cats. There is a nature walk with lakes and ponds, and there are cart rides. You can see a blacksmith and wheelwright at work, and relax in the licensed restaurant. There's a vast adventure playground with a terrifying twelve-metre free-fall slide. The owl sanctuary simulates moonlit conditions so you can see the barn owls flying about as they would in the countryside at dawn or dusk.

On-Site Activities: A one-hour show twice daily.

Educational Facilities: Education officer/centre; school visits.

The Monkey Sanctuary

Looe, Cornwall PL13 1NZ Tel: (0503) 262532

○   Open daily 10.30-5 May to Sep, Thurs to Sun in winter.

£   Adults £3.50, children £1.50, concession £2.50. Group rates (for schools): £2.50.

☞   From Liskeard or Plymouth take A38 and follow signs to B3253 at No Man's Land between East Looe and Hessenford.

Facilities: Car and coach parking. Toilets Refreshments Gift/book shop Play area.

Facilities for Disabled: Wheelchair access.

Restrictions: No dogs except in car park.

Description: The sanctuary was set up to rescue monkeys from zoos and as pets, and to set them in as natural an environment as possible, which it does well. The large enclosures contain Amazon woolly monkeys. In fact, this is the only place outside Amazonia where these monkeys breed. The monkeys, including mothers and babies, roam freely. You can see them showing all sorts of natural behaviour: playing, grooming, foraging for insects, leaves and nuts. They swing from tree to tree, often using their powerful prehensile tails.

Keepers will talk to visitors about the monkeys and displays tell you about their rainforest home and the threats to its future. Visitors are also given a chance to meet the monkeys when the mothers and babies are invited down. A Victorian garden is also open to the public, as well as a wildlife pond.

On-Site Activities: Talks given every hour.

Educational Facilities: Education officer/centre; school visits. Visits can be arranged to schools or the children can visit the sanctuary. Various workshops are held throughout the year. This is not a zoo; emphasis is given to education and observation.

Newquay Zoo

Trenance Leisure Park, Newquay, Cornwall Tel: (0637) 873342

○   Open daily 10-5 Easter to end Oct; last admissions at 4.
£   Adults £3.90, children £2.40, family £11.00, concession £1.60. Group rates: on application.
☞   Follow signs in Newquay town centre.

Facilities: 🅿 Car and coach parking 🚻 Toilets
   🍽 Refreshments 🛍 Gift/book shop 🛝 Play area.

Facilities for Disabled: ♿ Wheelchair access.

Restrictions: 🐕 No dogs.

Description: The zoo is part of the Trenance Leisure Park, so if the Water World does not exhaust you, a visit to the zoo could be in order. In the tropical house rare breeding and education centre, you can encounter many exotic insect and reptile species. The free-flight aviary has 100 birds set in a rainforest environment. At the children's zoo youngsters can mingle with the chickens, ducks and rabbits. Children can also use up excess energy on the assault course and in the activity play park. A giant-sized warren shows you what it is like to be a rabbit living underground. The aquarium, dragon maze and Tarzan trail will keep you busy, and look out for the lions, pumas, bears, deer, wallabies, camels, penguins, llamas and pythons. The café offers refreshments and there's a gift shop for souvenirs.

On-Site Activities: Tarzan trail; maze; animal encounters; Japanese gardens; play areas.

Educational Facilities: Education officer/centre; school visits.

Paradise Park

16 Trelissick Road, Hayle, Cornwall TR27 4RY Tel: (0736) 753365

- ○ Open daily 10-6 May to early Sep, 10-4 in winter.
- £ Adults £4.25, children £2.25, OAPs £3.75.
  Group rates: on application.
- ☞ On A30 near St Ives. Rail: Hayle station (½ mile).

Facilities: 🅿 Car and coach parking 🚻 Toilets – mother and baby facilities ☕ Refreshments 🛍 Gift/book shop 🛝 Play area.

Facilities for Disabled: ♿ Wheelchair access.

Restrictions: 🐕 No dogs.

Description: The park is home to the most outstanding collection of rare birds. It won an award from the Zoo Federation for the design of its parrot aviary which houses parrots from all over the world: hyacinth macaws, leadbeater cockatoos, eclectus parrots, Amazons and many others. Aviaries in the garden are home to a comprehensive collection of owls including Bengal eagle owls, snowy owls and boobook owls, many of which take part in regular flying displays. There is a specially designed barn where the light level is kept low so that barn owls can be seen at their most active. Closed-circuit televisions allow the visitors to watch the owls nesting and raising their chicks without disturbing them. Other exotic birds include Cuban flamingoes, toucans, white-backed vultures, many species of eagle, the very rare pink pigeon and Mauritius kestrels. Of particular interest to children are the penguin pool and otter enclosures, the mini-train and play area.

On-Site Activities: Free-flying bird display twice daily in summer. Otter and penguin feeding times daily. Guided tours sometimes available.

Special Events: Families' Day on last Sunday in July.

Educational Facilities: Education officer; school visits. Educational information about birds, animals and conservation available.

Tamar Otter Park and Wild Wood

North Petherwin, near Launceston, Cornwall PL15 8LW
Tel: (0566) 85646

- ○ Open daily 10.30-6 Easter to end Oct.
- £ Adults £3.00, children £1.50, OAPs £2.50.
  Group rates: adults £2.50, children £1.25.
- ☞ At North Petherwin 5 miles north-west of Launceston off the B3254 Bude road.

Facilities: 🅿 Car and coach parking 🚻 Toilets – disabled and mother and baby facilities ☕ Refreshments 👍 Gift/book shop.

Facilities for Disabled: ♿ Wheelchair access. Toilets.

Restrictions: 🐕 Guide dogs only.

Description: This branch of the Otter Trust is set in twenty acres of beautiful wooded valley with a stream where British otters are bred for release into the wild. The smaller, playful Asian short-clawed otters are also bred here as are owls and, in addition, three different species of free-ranging deer can be seen. The two lakes and waterfall feature a large collection of British and European waterfowl, as well as peacocks, dazzling golden pheasants and wallabies. Refreshments can be obtained from the tea room and you can make use of the visitor centre and buy souvenirs from the gift shop.

On-Site Activities: Talks to school visits and other groups. Otters fed twice daily; waterfowl fed daily; deer fed throughout the day.

Educational Facilities: Education centre; school visits.

DEVON
Buckfast Butterflies & The Dartmoor Otter Sanctuary

Buckfastleigh, Devon TQ11 0DZ Tel: (0364) 42916

○ Open daily 10-5.30 or dusk Mar to end Oct. Other times by special arrangement only.

£ Adults £3.25, children £2.00. Group rates: on application.

☞ Follow signs off A38 Exeter to Plymouth dual carriageway at Dart Bridge junction. Bus: No. 88 from Newton Abbot to Darbridge Road stop, Buckfastleigh.

Facilities: 🅿 Car and coach parking 🚻 Toilets – mother and baby facilities ☕ Refreshments 🎁 Gift/book shop 🛝 Play area.

Facilities for Disabled: ♿ Wheelchair access; elevated viewing positions for wheelchairs.

Restrictions: 🐕 Guide dogs only.

Description: The butterfly garden consists of a large landscaped tropical garden with ponds, waterfalls and bridges providing an ideal environment for exotic butterflies and moths. All stages of their life cycle can be seen and photographed in close-up. Other creatures, such as quail and terrapins, can also be found in this jungle setting.

The otter sanctuary has large landscaped enclosures for the resident otters with elevated viewing positions and an underwater viewing tunnel where they can be seen at play. An ingenious system of mirrors allows you to see the otters in their dens or 'holts' without disturbing them.

On-Site Activities: Guided tours; can be arranged in French, German or Spanish for pre-booked parties. Slide presentations and lectures/question and answer sessions. Otter feeding four times daily.

Educational Facilities: Education officer/centre; school visits. Lecture/classroom facilities available. Workbooks for visiting schools; slide presentations etc.

West Country – Devon   171

Canonteign Falls & Country Park
Near Chudleigh, Devon EX6 7NT   Tel: (0647) 52434

○   Open daily 10-5 Easter to end Oct, Sun only in winter.

£   Adults £3.00, children £2.00, OAPs £2.50.

☞   Follow signs to 3 miles off A38 at Chudleigh.

Facilities: 🅿 Car and coach parking   🚻 Toilets
   ☕ Refreshments   🛍 Gift/book shop   🛝 Play area.

Facilities for Disabled: ♿ Wheelchair access to facilities only.

Description: Canonteign is set in the heart of a private 100-acre estate and contains beautiful waterfalls, lakes and abundant wildlife. The Lady Exmoor Falls has a sheer drop of sixty-seven metres, making it the highest in England. Waterfowl and aquatic plants abound, and these can be seen on strolls along the nature trails. Children can be kept occupied on the junior assault course and in the play barn. They can meet the miniature ponies and everyone can enjoy the collection of farm machinery and learn about the history of mining in the Teign valley. Refreshments are available at the licensed restaurant as well as the all-weather barbecue, and traditional Devon cream teas are served in the tea rooms.

Educational Facilities: School visits. Nature interpretation centre for 1993.

Dartmoor National Park

The Dartmoor National Park Authority, Parke, Haytor Road, Bovey Tracey, Devon TQ13 9JQ Tel: (0626) 832093

Description: Dartmoor is one of the few wild places left in England. The moor lies on a huge granite batholith, relic of an ancient volcanic past. The craggy granite tors rise up out of the heather and blueberry moors. Rich boggy land adds to a mixture which harbours some unique and beautiful wildlife and plants. The moors and bogs can be treacherous, and the weather can change very quickly, blanketing the moors in mist even in midsummer. The famous wild ponies graze alongside deer, cattle and sheep. Thousands of years ago, many more people lived on Dartmoor and the remains of their granite huts, pounds and burial mounds can still be seen. There is a High Moorland Visitor centre (see page 173) which provides an interpretive experience of the landscape and includes audio-visual and interactive displays.

Exmoor Bird Gardens

South Stowford, Bratten Fleming, Barnstaple, Devon EX31 4SG Tel: (05983) 352

- ○ Open daily 10-6 Apr to Oct, 10-4 Nov to Mar. Closed Christmas Day.
- £ Adults £3.50, children £2.25. Group rates: adults £3.00, children £1.75.
- ☞ A399 (B3226). Bus: No. 310 Barnstaple to Lynton.

Facilities: ᴾ Car and coach parking ⋔ Toilets – disabled facilities ⚏ Refreshments ⛬ Gift/book shop ⛺ Play area.

Facilities for Disabled: ♿ Wheelchair access. Toilets.

Restrictions: ✘ No dogs or other pets.

Description: The twelve-and-a half-acre natural and formal bird gardens contain the largest collection of tropical birds in north Devon. A few examples especially worth looking for are the Humboldt penguins which have been hand-reared and are free to roam the park and mix with the visitors, as well as pelicans, hornbills, cockatoos and macaws. There are also lemurs, monkeys, llamas, maras and sika deer. Native wildfowl may also be seen in the garden area. A great deal of consideration has gone into the construction of all enclosures to ensure that, where possible, the animals and birds have a near-natural habitat. The success of this can be seen in their breeding programmes.

Educational Facilities: School visits. Staff give talks to local schools and institutions, taking birds and animals with them.

High Moorland Visitor Centre

The Ducky Hotel, Tavistock Road, Princetown, Yelverton, Plymouth, Devon PL20 6QF Tel: (0822) 89414

○ Open daily 10-5.

£ Admission free.

☞ Follow signs to Princetown and park in the main car park adjacent to the visitor centre. Bus: from Plymouth, Tavistock, Exeter. Telephone (0392) 38200 or (0752) 382800 for bus details.

Facilities: 🅿 Car and coach parking 🚻 Toilets
  ☕ Refreshments nearby 🛍 Gift/book shop.

Facilities for Disabled: ♿ Wheelchair access. Loop induction system.

Description: Opening in 1993, the High Moorland Visitor Centre aims to provide an interpretative experience of the unique Dartmoor landscape and will include audio-visual and interactive displays to help visitors enjoy Dartmoor National park.

On-Site Activities: Telephone for up-to-date details.

Educational Facilities: Education officer; school visits. Telephone for information on (0822) 89565.

The Milky Way and North Devon Bird of Prey Centre

Downland Farm, Clovelly, Devon EX39 5RY Tel: (0237) 431255

- ○ Open daily 10.30-6 Apr to end Oct; Nov to Mar by appointment.
- £ Adults £3.50, children £2.00.
  Group rates (over 20): 50p discount.
- ☞ On A39 Bideford to Bude road 2 miles from Clovelly. Bus: Devon Bus No. 319 Hartland to Bideford.

Facilities: 🅿 Car and coach parking 🚻 Toilets
🍴 Refreshments 🛍 Gift/book shop 🎪 Play area.

Facilities for Disabled: ♿ Wheelchair access.

Restrictions: 🐕 Dogs on leads. No dogs in falconry mews and viewing galley of milking parlour.

Description: The Milky Way is a 205-acre family-run working dairy farm with some two and a half acres under cover set in ten acres of adventure playground, picnic area and display arena. The old farmyard now acts as the café, serving teas. The magnificent purpose-built North Devon Bird of Prey Centre includes a barn owl breeding aviary for conservation work and a bird hospital. There are two flying displays a day, using birds from the collection including a European eagle owl, lanner and lugger falcons, harris and ferruginous hawks, common and red-tailed buzzards, kestrels and barn owls. The countryside collection contains a farming family's history of more than 900 items, including a hand-carved caravan, and handcarts.

On the farm, you can watch the herd being milked, and calves being born in spring. Lambs, goat kids and chicks add to the ranks of baby animals, some of which visitors can bottle-feed. You can buy some of the end products of the farm in the shop as well as local crafts such as ceramics from the Raku pottery which you can watch being made.

On-Site Activities: Twice-daily flying displays of birds of prey and bottle feeding. Hand-milking; working pottery; environmentally friendly laser clay pigeon shooting; cuddle baby animals. All involve audience participation.

Educational Facilities: School visits.

Miniature Pony Centre

Wormhill Farm, North Bovey, near Newton Abbot, Devon TQ13 8RG
Tel: (0647) 432400

- ○ Open daily 10-5 throughout the year; last admissions at 4.
- £ Adults £3.95, children £2.95, family £1.00 discount. Group rates: 25p discount; schools, playgroups etc. £2.50.
- ☞ On B3212, 3 miles west of Moreton Hampstead. Bus: Transmoor link Exeter to Plymouth during summer.

Facilities: 🅿 Car and coach parking ♯♯ Toilets – disabled and mother and baby facilities ☕ Refreshments 👍 Gift/book shop 🛝 Play area.

Facilities for Disabled: ♿ Wheelchair access. Toilets.

Restrictions: 🐕 No dogs in centre. Kennels available free of charge.

Description: Set amidst the natural beauty of Dartmoor National Park, the centre is home to a number of miniature animals based on a working stud of miniature Shetland ponies. For twenty years the stud has been breeding miniatures with about forty foals born each year. Miniature donkeys, pygmy goats, pigs and Dexter cattle are bred here and the visitor is encouraged to approach and touch these animals in their paddocks or stables. Netherland dwarf rabbits and chinchillas can also be handled. Several lakes provide homes and resting sites for a variety of waterfowl including Canada geese, moorhens and ornamental waterfowl. A collection of unusual bantams can also be found.

Educational Facilities: School visits.

National Shire Horse Centre

Yealmpton, Plymouth, Devon PL8 2EL Tel: (0752) 880268

○ Open daily 10-5 throughout the year except Christmas Eve, Christmas Day and Boxing Day.

£ Adults £4.95, children £3.30, family £15.50, OAPs £4.45. Group rates: £3.95.

☞ A38 to Plymouth then follow signs. Situated on A379 Plymouth to Kingsbridge road.
Bus: No. 93 from Plymouth.

Facilities: ᴾ Car and coach parking ⚥ Toilets – mother and baby facilities ☕ Refreshments 🛍 Gift/book shop ⛰ Play area.

Facilities for Disabled: ♿ Wheelchair access.

Description: There are over thirty Shire horses on display at this 100-acre Devon farm, which features regular parades every day of the 'gentle giants'. There are also flying displays at the falconry centre which has a Russian Steppes eagle, hawks, buzzards, owls and other birds of prey. There is a butterfly house to visit, and a pets area with lambs, pigs, goats, calves and rabbits. A new Environmental Walk goes through fields and woods by the River Yealm.

On-Site Activities: Parades of horses three times a day. Falconry flying displays twice a day.

Special Events: Special events, many at no extra cost, include teddy bears' picnic, dollies' tea party, Western weekend, steam and vintage rally.

Educational Facilities: School visits.

Paignton Zoo

Totnes Road, Paignton, Devon TQ4 7EU Tel: (0803) 527936

○ Open daily 10-6 in summer, 10-5 in winter. Closed Christmas Day.

£ Adults £5.40, children (3+) £3.30, family (2 + 2) £15.30.

☞ 400 metres from A380 Torbay ring road at Totnes (A385) junction; 1 mile from Paignton town centre.

Facilities: 🅿 Car and coach parking 🚻 Toilets – mother and baby facilities ☕ Refreshments 🛍 Gift/book shop 🎠 Play area.

Facilities for Disabled: ♿ Limited wheelchair access. Some Braille literature.

Restrictions: 🐕 Guide dogs only.

Description: This is one of England's largest zoos with over 1,300 animals, many of which are endangered and involved in captive breeding programmes. There are big cats, elephants, giraffes, monkeys, parrots and an assortment of reptiles. The zoo is set in seventy acres of lovely botanical gardens. The ark family activity centre provides hours of fun and learning for the whole family. There is also a miniature railway and super-x simulator. This zoo mixes all the fun of meeting the animals as well as communicating the conservation message in an entertaining way.

On-Site Activities: Talks; meet the keepers at feeding times Easter to end September.

Special Events: Full calendar of events such as reindeer at Christmas, summer art courses, fun days.

Educational Facilities: Education officer/centre; school visits.

Parke Rare Breeds Farm

Parke Estate, Bovey Tracey, Devon TQ13 9JQ Tel: (0626) 833909

○ Open Mon to Fri 10-6 and Sat and Sun 11-4 Easter to end Oct; last admissions 1 hour earlier.

£ Adults £3.50, children £1.75.
Group rates: on application.

☞ Take A382 to Bovey Tracey then B3387 towards Haytor. Turn into estate 200 metres from roundabout following signs. Bus: to Bovey Tracey (½ mile).

Facilities: P Car and coach parking Toilets Refreshments Gift/book shop Play area.

Restrictions: No dogs on farm. Dogs on leads allowed around free estate walks.

Description: This is an approved centre for rare breeds of farm animals and fowl; many here used to be farmed as long ago as the middle ages. Such centres help to breed and maintain these ancient bloodlines. The farm is set in beautiful parkland near Dartmoor. There are over seventy breeds of animal on display. There is a pit pony feature, a bunny village, 'Martin' vintage agricultural tool display, and a walk-in pets corner where children can meet and stroke the animals. Guided walks are available if booked in advance. Added facilities include a restaurant, craft centre, interpretation centre and the largest model pig display in the UK.

On-Site Activities: Guided walks can be pre-booked. Pit pony feature; bunny village; 'Martin' vintage agricultural tool display; walk-in pets corner.

Educational Facilities: Education centre; school visits.

Plymouth Aquarium

Citadel Hill, Plymouth, Devon PL1 2PB Tel: (0752) 222772

○ Open daily 10-6 Apr to Sep, 10-5 Oct to Mar.
£ Adults £2.00, children £1.00, family £5.00.
☞ A38 to Plymouth then follow signs to The Hoe.
Rail/Bus: to Plymouth then bus to The Hoe.

Facilities: ♯ Toilets ⛾ Refreshments ♣ Gift/book shop.
Facilities for Disabled: ♿ Wheelchair access.
Restrictions: ✘ No dogs.
Description: This aquarium sets out to educate its visitors about the wealth of marine creatures, together with the problems they face in the wild. Marine creatures are set in attractive displays. There is a whole range of wildlife, much of it native to our shores. You can see crustaceans such as lobsters and crabs, also octopuses and more than fifty different species of fish. Further displays explain shoreline conservation and show you what you might find if you go beachcombing. It also explains the ecology and life of shore creatures and plants. A good way of seeing the wildlife found off the shore of Plymouth without actually having to get wet and cold!
Educational Facilities: School visits.

Quince Honey Farm

North Road, South Molton, Devon EX36 3AZ Tel: (0769) 572401

○ Open daily 9-6 Easter to end Sep, 9-5 Oct.
Shop open 9-5 in winter.
£ Adults £2.75, children £1.35. Group rates: £1.60.
Special rates for school and disabled groups.
☞ Follow signs from A361 in South Molton.
Bus: to South Molton from Barnstaple.

Facilities: 🅿 Car and coach parking ♯ Toilets
⛾ Refreshments ♣ Gift/book shop.
Facilities for Disabled: ♿ Wheelchair access.
Special rates for disabled groups.

Description: The Quince Honey Farm houses a splendid exhibition of living honey bees. All the hives are glass-sided so visitors can view the bees at work in complete safety. The unique opening hives are an entertaining education for young and old alike. The live action of the busy bees is complemented with posters, photographs and video shows. You can see the bees building the waxen cells and making honey. If you are lucky you may spot the queen bee and watch the workers doing the waggle dance which tells the other bees where to look for flowers containing nectar. Between November and Easter the bees are not active, but the shop is open.

Educational Facilities: School visits.

Shaldon Wildlife Trust

Ness Drive, Shaldon, Devon TQ14 0HP Tel: (0626) 872234

- ○ Open daily 10-6 in summer, 11-4 in winter.
- £ Adults £2.30, children £1.45, family £6.00. Group rates: 15% discount.
- ☞ Off B3188 Torquay to Teignmouth road. Bus: No. 85 to Ness then walk (¼ mile).

Facilities: 🅿 Car and coach parking 🚻 Toilets 🛍 Gift/book shop.

Restrictions: 🐕 No dogs.

Description: The Shaldon Wildlife Trust is set amongst woodland and boasts some of the rarest and most unusual small animals in the world. All the animals here are captive-born and are part of a breeding programme designed to stock other zoos and reintroduce creatures into the wild, hence there are always babies to be seen here. Especially cute are the tiny threatened monkeys, the marmosets and tamarins, but equally fascinating are the owls, reptiles, amphibians, tarantulas, agoutis, saki, Diana and squirrel monkeys, meerkats and mongooses! A very enjoyable way of seeing conservation at work.

On-Site Activities: Talks; guides.

Educational Facilities: Education officer; school visits.

Woodland Leisure Park

Blackawton, Totnes, Dartmouth, Devon TW9 7DQ   Tel: (080 421) 598/680

○ Open daily throughout the year; shop and café closed weekdays Nov to Mar.

£ Adults £3.50, children £2.50, family (2 + 2) £11.00. Group rates and concessions: on application.

☞ From A38 at Buckfastleigh turn right to Totnes then right on A3122 to Kingsbridge. At Halwell turn right on A3122 to Dartmouth and park is 2½ miles on left.

Facilities: 🅿 Car and coach parking 🚻 Toilets – mother and baby facilities ☕ Refreshments 🛍 Gift/book shop 🛝 Play area.

Facilities for Disabled: ♿ Wheelchair access.

Restrictions: 🐕 Dogs on leads.

Description: The park is set in sixty acres with thirty-two acres of ancient woodland and monastic ponds, built to provide food for the monks who used to live here. An extensive collection of waterfowl now flourish on the ponds. The honey farm features live bee chambers and displays. There is a large animal farm complex with sheep, ducks and geese, amongst others. Visitors can also meet the Shire horses in their meadow. The badger set is home to sixteen of these native carnivores. There is a hide where you can catch a glimpse of sika deer. The monks' nature trail will introduce visitors to a variety of native wildlife, as well as showing you the monks' cave and haunted pool. The more energetic can have fun on the venture activities such as the commando assault course found within the woodland, and woodland trails.

On-Site Activities: Bee talk and demonstration by bee keeper.

Special Events: Live entertainment days throughout the school holidays.

Educational Facilities: Education officer; school visits. Educational packs available on flowers, monks' nature trail, birds' eggs, the busy bee. School camp facility.

SOMERSET
Brean Down Bird Garden

Burnham-on-Sea, Somerset TA8 2RS Tel: (0278) 751209

○   Open daily 9-6 Mar to end Oct; 10-5 in winter, weather permitting.

£   Adults £1.95, children 95p. Group rates: 20% discount.

☞   From Burnham take cost road to Berrow then Brean to Brean Down. Bus: from Weston or Burnham-on-Sea.

Facilities: Car and coach parking   Toilets – mother and baby facilities   Refreshments   Gift/book shop.

Facilities for Disabled: Wheelchair access.

Restrictions: Dogs on leads.

Description: This bird garden is only a few miles away from one of the West's most famous seaside resorts, Weston-super-Mare, and close to a National Trust hill where annual migration counts take place. The birds are kept in aviaries in a garden which is set in an acre of its own land. The bird garden boasts one of the largest collections of Australian parakeets in the country. Some are exceptionally rare. Many other tropical birds can be seen, including zebra finches, Amazon cockatoos, African greys and scarlet macaws. During the summer months a variety of baby birds which are bred on site can be seen on request.

On-Site Activities: Talks can be pre-booked.

Educational Facilities: School visits.

Cricket St Thomas Wildlife Park

Near Chard, Somerset TA20 4DD Tel: (0460) 30755

○ Open daily 10-6 in summer, 10-dusk in winter.

£ Adults £5.50, children £3.50, OAPs £4.50.
Group rates: adults £3.50, children £2.50, OAPs £3.00 (one person free in 20).

☞ From M5 junction 25, follow signs to A30 between Char and Crewkerne.

Facilities: 🅿 Car and coach parking 🚻 Toilets – mother and baby facilities ☕ Refreshments 🛍 Gift/book shop 🎠 Play area.

Facilities for Disabled: ♿ Wheelchair access.

Restrictions: 🐕 Dogs on leads.

Description: The sheltered valley in which the park is situated provides a superb habitat for a great many of the animals and birds to be seen, many of which roam free. Around the lakeside there are waterfowl and flamingoes to enjoy but also deer, zebras, camels, llamas and wallabies can be seen at close range. Keep an eye out for the Asian elephants, they may come to you rather than you going to them if you are lucky enough to meet them on a daily walk around the estate. There are jaguars, leopards, servals and lynxes to be found and a lemur house where you can be entertained by the energetic behaviour of the monkeys and marmosets as well. The sealions will delight you with their antics, performing throughout the summer. Look out for the landscaped walk-through aviary where you can get an idea of what it is like to take a stroll in a tropical habitat and see an interesting display of birds that are found in such places. Visit the Heavy Horse Centre, or wander around the sixteen acres of beautiful gardens.

On-Site Activities: Talks available on request.

Special Events: Custom and American car show in early May. Heavy horse show and steam rally in late May. Game and country fair in late June. Classic car day in mid-July.

Educational Facilities: Education officer/centre; school visits.

Exmoor National Park

Exmoor National Park Authority, Exmoor House, Dulverton, Somerset TA22 9HL Tel: (0398) 23665

Description: At 267 square miles, Exmoor is one of our smaller National Parks. Straddling the boundaries of west Somerset and north Devon, the park towers over the Bristol Channel, its magnificent coastline giving superb views across to the distant Welsh mountains. Inland, wild heather moorland gives way to gentle valleys and farms and to the east the Brendon Hills provide a forested contrast.

Exmoor can boast the largest wild herd of red deer outside Scotland. In early autumn it is possible to observe Britain's largest wild animal actively rutting. Two pure-bred herds of Exmoor pony, exclusive to the park, roam the moorland. Other animals to be spotted include the buzzard and grey squirrel. Ideal for walking or riding, Exmoor provides a tranquil, second-to-none experience.

Tropiquaria

Washford Cross, Watchet, Somerset TA23 0JX Tel: (0984) 40688

○ Open daily 10-6 Mar to Oct, 11-4 weekends and holidays in Nov, Jan, Feb and 28-31 Dec.

£ Adults £2.90, children £1.50, concession £2.40. Group rates: adults £2.40, children £1.30.

☞ From M5 junction 23 to Bridgwater, follow A39 to 1½ miles west of Williton on Minehead road. Bus: from Williton, Watchet or Minehead. Rail: West Somerset railway, Watchet and Washford station (1½ miles).

Facilities: 🅿 Car and coach parking 🚻 Toilets 🍴 Refreshments 🛍 Gift/book shop 🛝 Play area.

Facilities for Disabled: ♿ Limited wheelchair access.

Restrictions: 🐕 Guide dogs only.

Description: Mostly an indoor attraction, Tropiquaria offers visitors the chance to walk through a tropical jungle and see colourful rainforest birds flying overhead or find others picking their way through the luxuriant jungle plants at ground level. There is the opportunity to meet a snake or to learn of the weird and wonderful reptiles that lurk in these warm, wet habitats. Look out for tamarins and tortoises, snakes and spiders, and if you dare, there is a creepy-crawly crypt to explore. Also downstairs is a stunning collection of exotic fish and a venture outside will bring you to aviaries and landscaped gardens that are home to lemurs and coatis. The added attractions include an adventure playground and the Shadowstring Puppet Theatre – this is well worth a visit for young and old alike and is at no extra charge.

On-Site Activities: Guided tours for groups. Puppet theatre. Holding and stroking a snake.

Educational Facilities: School visits.

WILTSHIRE
Cholderton Rare Breeds Farm and Gardens

Amesbury Road, Cholderton, Salisbury, Wiltshire SP4 0EW
Tel: (0980) 64438

- ○ Open daily Easter to end Oct.
- £ Adults £3.00, children £1.50, OAPs £2.50.
  Group rates: 10% discount if pre-booked.
- ☞ Follow signs from A338 Salisbury to Marlborough road or A303 Andover to Amesbury road.
  Bus: Nos. 63, 64 Salisbury to Cholderton.

Facilities: Car and coach parking Toilets Refreshments Gift/book shop Play area.

Facilities for Disabled: Wheelchair access.

Restrictions: Dogs on leads. Children must be accompanied and supervised by an adult.

Description: The farm covers over fifty acres and straddles the Hampshire/Wiltshire border, enjoying views across the valley as far as Salisbury Cathedral spire some ten miles away. The farm prides itself on a wide variety of British breeds and farm animals which form part of our national heritage and are close to extinction. Within the farm there is a breeding centre for Castlemilk Mourit, Norfolk Horn, Portland and Hebridean sheep, as well as Kerry cows, Bagot and golden Guernsey goats. The rabbit world exhibits over fifty breeds. These and friendly pygmy goats can be found in the woodland pets park. The visitor may have time to identify the animals on the footprint trail or walk along the nature track hoping to spot deer, rabbits, squirrels and other wildlife.

On-Site Activities: Nature trail through woodland to badger sett. Group guides and talks by arrangement.

Educational Facilities: Education officer/centre; school visits.

Farmer Giles Farmstead

Teffont, Salisbury, Wiltshire SP3 5QY Tel: (0722) 716338

○ Open daily 10.30-6 Feb to early Nov and every weekend until Christmas. Pre-booked parties throughout the year.

£ Adults £3.00, children £2.00, OAPs £2.50.
Group rates: 20% discount.

☞ From Salisbury via A20 and B3089, about 11 miles southwest of Stonehenge on A303 London to Exeter road. Bus: from Salisbury to Teffont.

Facilities: 🅿 Car and coach parking 🚻 Toilets 🍽 Refreshments 🎁 Gift/book shop 🎠 Play area.

Facilities for Disabled: ♿ Wheelchair access; over 20,000 sq ft (1860 sq metres) under cover and paved paths to and around pond are ideal for wheelchairs. Trails for visually handicapped.

Restrictions: 🐕 Dogs welcome.

Description: The story of our daily pinta begins each afternoon as 150 Friesian cows wander in from their pastures to the milking parlour. Milking can be viewed from a gallery where there is a display about dairying through the ages. Elsewhere on the farm, visitors can bottle-feed lambs, as well as meet and touch rabbits, goats, ponies, pigs, calves, ducks, geese, guinea fowl and bantams. From the nature walk, pheasants, partridges and hares can be seen, and more occasionally, deer, stoats, owls, lambs and herons. The farm also has rare breeds and a display about farming through the ages.

On-Site Activities: Beech belt nature walk; guides and talks available.

Special Events: Military vehicle weekend early May; farm working weekend early June; charity weekend end July; country crafts weekend in September.

Educational Facilities: Education officer/centre; school visits. Educational pack available.

Lions of Longleat Safari Park

Warminster, Wiltshire BA12 7NJ Tel: (0985) 844328

- ○ Open daily 10-6 early Mar to end Oct. Closing times may vary in Mar and Oct.
- £ Adults £5.00, children £3.50, OAPs £4.00. Group rates: on application.
- ☞ On A362 Warminster to Frome road ½ mile off the new Warminster bypass on the A36 Bath to Salisbury trunk road. M4 junction 17 or 18. From London M3, A303, A36. From Devon M5 junction 25, A361 to Frome. Bus: service to main gate, telephone Badgerline (0225) 64446 for details. Rail: Warminster, Frome, Westbury stations (3 miles from each).

Facilities: 🅿 Car and coach parking 🚻 Toilets – disabled and mother and baby facilities 🍽 Refreshments 🎁 Gift/book shop ⛱ Play area.

Facilities for Disabled: ♿ Wheelchair access. Toilets.

Restrictions: 🐕 No dogs in maze, adventure castle, butterfly garden and animal reserves. Free kennelling only while in safari park. No soft-top cars.

Description: Set in rolling parkland, the sixteenth-century mansion lords over a 300-acre drive-through safari park. There you can enjoy the white tigers, monkeys, rhinos, giraffes, camels, buffaloes, elephants and, of course, lions! Children can get even closer to the animals at pets corner. Youngsters can also get rid of any excess energy in the two-acre playpark; if this fails then the Dr Who exhibition and the world's largest maze should do the trick. But the fun doesn't stop there. Visitors can also explore the Victorian kitchens and doll's house collection; ride on the train and safari boat to see the gorillas, hippos and sealions; and wonder at Lord Bath's vehicle collection. If you aren't totally exhausted by then, the simulator will certainly finish you off. This allows you to ski at Val d'Isére, or dog-fight with the Red Baron without ever leaving the simulator capsule. Much-needed refreshments can be obtained from the restaurant, pub or café.

On-Site Activities: Safari trail; maze; adventure castle; butterfly garden; doll's houses; Dr Who exhibition.

Educational Facilities: School visits.

Woodland Park and Heritage Museum

Brokerswood, near Westbury, Wiltshire Tel: (0373) 822238

○ Open daily 10-dusk throughout the year.

£ Adults £2.00, accompanied children under 14 free of charge, OAPs £1.75. Group rates: on application.

☞ Follow signs off A36 Warminster to Bath road at Standerwick.

Facilities: Car and coach parking   Toilets – mother and baby facilities   Refreshments   Gift/book shop   Play area.

Facilities for Disabled: Wheelchair access.

Restrictions: Dogs on leads.

Description: The estate comprises eighty acres of working woodland, ensuring the survival of a wide range of trees, plants and animals. Major features at the park include a wildfowl lake and many nature trails. A guided walk around the otherwise private areas of the estate not normally open to the public provides the visitor with an opportunity to hear how the park has been developed over the past thirty years. Other attractions include the Smoky Oak Railway covering one-third of a mile. This operates at weekends, school holidays and public holidays in the summer, or for groups or parties throughout the year by arrangement. There is also a heritage museum which encourages visitors to participate and learn about conservation. Educational programmes are also available to schools, including an on-site laboratory.

On-Site Activities: Guided walks; suppers; children's activity trails by arrangement.

Special Events: Calendar of special events available on request.

Educational Facilities: School visits.

YORKSHIRE AND HUMBERSIDE

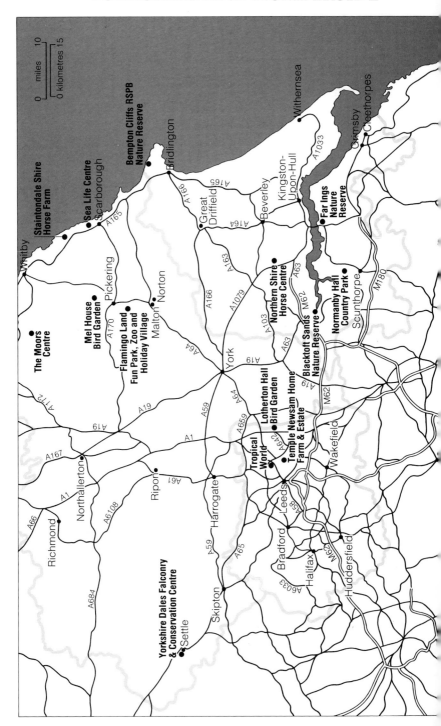

HUMBERSIDE
Bempton Cliffs RSPB Nature Reserve

Bempton, near Bridlington, Humberside Tel: (0262) 851179

○ Open daily 10-5 Easter to end Sep.

£ Admission free. Car park: cars £1.00, minibus £2.50, coach £5.00, RSPB members free.

☞ Reserve is 1 mile north of Bempton village, which is 3 miles north of Bridlington and 12 miles south of Filey. From Bempton, turn north at White Horse pub to car park 1 mile up road near cliff top. Rail: Bempton station.

Facilities: ▣ Car and coach parking ⚦ Toilets ▣ Refreshments ♣ Gift shop.

Facilities for Disabled: ♿ Wheelchair access.

Restrictions: 🐕 Dogs on leads. Cliffs can be dangerous.

Description: With chalk cliffs rising to 400 feet and extending for three and a half miles along the coast, this reserve contains England's largest seabird colony which is active from April to the end of August or early September. Eight species of seabird breed here: gannets, kittiwakes, fulmars, herring gulls, shags, guillemots, razorbills and puffins. Other birds to see include corn buntings, meadow and rock pipits, tree sparrows and migrant birds in season. The puffins and gannets are the most popular with visitors and the best time to see them is in May, June and July. There are four staff on site, five viewpoints and an interpretation centre.

On-Site Activities: Guided walks and open days; special escorted walks and talks can be arranged by writing to 11 Carlton Avenue, Hornsea, East Yorkshire HU18 1JG.

Educational Facilities: Education officer; school visits. An educational programme is run from April to July. Schools book through RSPB Lincoln office, The Lawn, Union Road, Lincoln NL1 3BU, (0522) 535596. Each session lasts two hours and is linked to the national curriculum.

Blacktoft Sands Nature Reserve

Ousefleet, near Goole, North Humberside Tel: (0405) 704294

- ○ Open Wed to Mon 9-9 Apr to Aug, 9-dusk Sep to Mar.
- £ Adults £2.50, children 50p. RSPB and YOC members free of charge.
- ☞ A161 from Goole to Swinefleet then unclassified road to car park ½ mile beyond Ousefleet.
  Bus: from Goole, 2-3 hourly service (not Sun).

Facilities: 🅿 Car and coach parking 🚻 Toilets 🛝 Play area.

Facilities for Disabled: ♿ Wheelchair access.

Restrictions: 🐕 No dogs.

Description: This nature reserve includes a nature trail with six observation hides overlooking lagoons and reedbeds where visitors can watch the local bird life. Species to look out for include marsh harriers, bearded tits and avocets in summer, hen harriers in winter, and many passing birds on their spring and autumn migrations. Butterflies and dragonflies are plentiful in summer, and foxes can often been seen on the reserve.

On-Site Activities: Guided walks. Free open day in June each year.

Educational Facilities: School visits. Pond dipping and school projects by arrangement.

Far Ings Nature Reserve

Far Ing Lane, Barton-upon-Humber, South Humberside DN18 5RG
Tel: (0652) 34507

○ Reserve open daily throughout the year; visitor centre open Wed, Sat, Sun Apr to Oct.

£ Admission free.

☞ Follow signs west of Humber Bridge out from Barton-upon-Humber.

Facilities: Car and coach parking (coaches by arrangement) Toilets Refreshments on request Gift/book shop.

Facilities for Disabled: Limited wheelchair access.

Restrictions: No dogs.

Description: Set alongside the Humber bank, the nature reserve encompasses four different habitats. Flood clay pits now known as the Barton reedbeds stretch for 100 acres and provide an exclusive retreat for several rare breeds of bird and moth (over 200 species of moth alone). The reeds, when mown, are sold for thatch. The foreshore includes an estuarine saltmarsh and is home to a number of specialised birds such as the endangered bearded tit and the bittern. A large number of reed warblers, water rails and ducks can be spotted. Areas of scrub provide cover for a variety of birds including white throats. Meadows are rich with wild flowers. There are seven hides and an extensive network of footpaths for easy walking. A guided tour can be included.

On-Site Activities: Self-guided nature trails. Walks and talks by arrangement.

Special Events: Various special events throughout the year. Leaflet published each year.

Educational Facilities: Education officer/centre; school visits by arrangement.

Normanby Hall Country Park

Normanby, Scunthorpe, South Humberside DN15 9HU Tel: (0724) 720588

○ Park open daily 9-8 or dusk throughout the year. Hall and farming museum open afternoons Apr to Sep.

£ Admission free. Car parking charge 60p weekdays, £1.50 weekends.

☞ Off B1430 4 miles north of Scunthorpe.

Facilities: 🅿 Car and coach parking 🚻 Toilets – mother and baby facilities ☕ Refreshments 🛍 Gift/book shop 🛝 Play area.

Facilities for Disabled: ♿ Wheelchair access.

Restrictions: 🐕 Dogs on leads.

Description: The park covers an area of 350 acres. The traditional parkland comprises woodland and a deer park with two deer herds. There is an abundance of native wildlife that can regularly be seen or heard. There are two species of woodpecker found here as well as birds of prey like tawny owls, sparrowhawks and kestrels. Flocks of tits and warblers are a noisy, colourful feature. In summer, migrants such as six species of warbler come here to breed. Nature trails help visitors to explore the woods and deer park. The regency mansion is well worth a visit, as is the farming museum showing examples of ways the countryside was managed in years gone by.

On-Site Activities: Special events every Sunday during the season.

Educational Facilities: Education officer/centre; school visits.

NORTH YORKSHIRE
Flamingo Land Fun Park, Zoo and Holiday Village

Kirby Misperton, Malton, North Yorkshire YO17 0UX Tel: (0653) 86287

○ Open daily 10-5 weekdays, 10-6 weekends Easter to end Oct.

£ £7.00 (children under 4 free). Group rates: on application.

☞ Off A169 Malton to Pickering road.

Facilities: ⓟ Car and coach parking ♦♦ Toilets – mother and baby facilities ☕ Refreshments 🛍 Gift/book shop ⛹ Play area.

Facilities for Disabled: ♿ Wheelchair access.

Description: This is Europe's largest privately owned zoo with over 1,000 animals, birds and reptiles in natural habitat surroundings. Big cats include the lion, tiger, leopard, puma and lynx. There are a number of other large mammals such as elephants, wolves, camels and a breeding group of bison. Dolphin, sealion and parrot shows regularly take place throughout the day. An aquarium displays a variety of fish and a reptile house accommodates pythons, alligators and a dwarf crocodile. The zoo also has a number of birds of prey.

The Funpark and Holiday Village provide hundreds of attractions, breathtaking rides and slides, including the North's biggest corkscrew, roller coaster, thunder mountain, flying carpet, log flume, circus and Cinema 180.

Educational Facilities: School visits.

Mel House Bird Garden

Mel House, Newton-on-Rawcliffe, Pickering, North Yorkshire YO18 8QA
Tel: (0751) 76538

- ○ Open daily 1-5 Apr to end Sep.
- £ Adults £1.50, children £1.00.
- ☞ From Pickering to Newton-on-Rawcliffe village centre.

Facilities: 🅿 Car parking 🚻 Toilets ☕ Refreshments 🎁 Gift/book shop 🛝 Play area.

Facilities for Disabled: ♿ Wheelchair access; designed with the disabled in mind.

Restrictions: 🐕 Dogs on leads. Children under 14 must be accompanied by an adult.

Description: Set amidst three acres of natural gardens, Mel House Bird Garden is a working wildlife sanctuary and specialises in owls. Many British species of owl are kept in the large aviaries and a flying display is performed daily. Other owls include the African spotted, Bengal eagle and barn owl; a kestrel is also kept. Members of the public are given the opportunity to handle these incredible birds. Educational talks and lectures are provided. A number of hand-reared rare breed farm animals can be hand-fed. These include sheep, pigs, goats, poultry and pheasants. A teashop is on the premises. Sick or unwanted animals are donated to the sanctuary and are reintroduced into the wild or given to a wildlife park.

This is a non-profit-making organisation. All monies raised go directly back to help wildlife.

On-Site Activities: Free-flying bird of prey display daily, including an educational talk.

Educational Facilities: School visits.

The Moors Centre

Danby Lodge, Lodge Lane, Danby, Whitby, North Yorkshire YO21 2NB
Tel: (0287) 660540

○ Open daily 10-5 Apr to end Oct, Sat and Sun 11-4 Jan to end Mar and Nov to end Dec.

£ Admission free.

☞ From A171 Teesside to Whitby road 8 miles east of Guisborough and 15 miles west of Whitby, turn off main road and follow signs (3½ miles).

Facilities: P Car and coach parking Toilets Refreshments Gift/book shop Play area.

Facilities for Disabled: Wheelchair access.

Restrictions: No dogs.

Description: A former shooting lodge has been adapted as a visitor centre for the North York Moors National Park. The grounds extend to thirteen acres including garden, woodland and riverside meadows. A special wild flower garden was established in 1992 showing the different habitats of the National Park with the appropriate wild flowers. It includes the three types of heather – bell, cross-leaves heath and ling – and woodland flowers such as herb Robert, golden saxifrage and red campion. A well-marked trail leads through the woods where you may see grey squirrels, blue tits and other woodland birds. By the river, you may be able to spot kingfishers or dippers.

On-Site Activities: An exhibition on how people have altered and affected the moorland landscape. Talks can be arranged; contact the centre for further details.

Special Events: In summer there is a programme of special events such as kite-making, wild flower talks and craft demonstrations of dry stone walling and spinning.

Educational Facilities: Education officer/centre; school visits must be pre-booked.

Northern Shire Horse Centre

Flower Hill Farm, North Newbald, York YO4 3TG Tel: (0430) 827270

- ○ Open Sun to Thurs 10-5 Easter to end Sep.
- £ Adults £2.00, children and OAPs £1.45.
  Group rates (over 20): 20% discount.
- ☞ Follow signs from A103. Grid reference 935384, Market Weighton OS map.

Facilities: 🅿 Car and coach parking 🚻 Toilets – disabled and mother and baby facilities ☕ Refreshments 👍 Small gift/book shop 🅿 Play area.

Facilities for Disabled: ♿ Wheelchair access. Toilets. Currently fundraising in order to provide a video room and better facilities for the disabled.

Description: The farm combines the best of the new and the old: the modern farm machinery is found side by side with the magnificent Shires that would have powered the farm in years gone by. Besides these largest of horses, visitors can meet other unusual farm animals, such as longhorn cattle, and see a collection of Victorian and Edwardian clothing. Smaller horses for riding and driving can be found in the stables, too. As well as the abundant native birds, you will also find exotic peafowl and guinea fowl. The farm museum displays farming bygones, machinery and a Victorian farmhouse kitchen. The harness room houses all the leather and brass tack for the heavy horses. The forge is still in full working order, and indeed is used to shoe the farm's horses. Refreshments are available at the tea rooms.

Educational Facilities: School, playgroup and group visits all have guided tours. Suitable for groups of children with learning difficulties.

Sea Life Centre, Scarborough

Scalby Mills Road, Scarborough, North Yorkshire YO12 6RP
Tel: (0723) 376125

○ From 10 daily except Christmas Day (late opening in summer).

£ Adults £3.95, children and OAPs £2.95, disabled £2.50. Group rates (10 or more): adults £3.10, children £2.10.

☞ Follow signs to North Bay Leisure Parks on Whitby Road beyond Watersplash and Kinderland. Bus: No. 110 shuttle bus from town centre all year; open-top bus along seafront to the centre in summer.

Facilities: ᴾ Car and coach parking ⚦ Toilets – mother and baby room ☕ Refreshments 🛍 Gift/book shop.

Facilities for Disabled: ♿ Wheelchair access.

Restrictions: 🐕 Guide dogs only.

Description: Scarborough Sea Life Centre takes visitors beneath the crashing waves and provides a fascinating glimpse into the Yorkshire coastline. From the depths of a spectacular underwater tunnel, visitors come face to face with hundreds of fascinating and exciting sea creatures from around our shores. There are shrimps, starfish, sharks, stingrays, conger eels and octopuses, to name just a few, and all can be seen from astonishing angles. There are regular talks and feeding demonstrations throughout the day. Other exhibits include a sea laboratory where marine biologists use state-of-the-art technology to bring the wonders of the sea a little closer.

On-Site Activities: Regular talks and feeding demonstrations throughout the day.

Educational Facilities: Educational officer/centre; school visits.

Staintondale Shire Horse Farm

Staintondale, Scarborough, North Yorkshire YO13 0EY Tel: (0723) 870458

- ○ Open Sun, Tues, Wed, Fri and bank holiday Mon 10.30-4.30 Easter to end Sep.
- £ Adults £2.50, children £1.50 (over 2 children, 50p each), OAPs £2.00. Group rates: 10% discount.
- ☞ Signposted from A171 Whitby to Scarborough road. Bus: from Scarborough, but 1 mile walk to farm.

Facilities: P Car parking ♦♦ Toilets ☕ Refreshments ♣ Gift/book shop ⚑ Play area.

Facilities for Disabled: ♿ Wheelchair access; farm, talks, demonstrations, video room and special refreshment area all accessible to wheelchairs.

Restrictions: 🐕 Dogs restricted to large grass car park.

Description: Visitors to the farm are encouraged to meet and stroke the Shire horses and to watch them at work and at play. You can also feed the ducks and hens. There are stables, the blacksmith's shop and a collection of farming bygones to explore. Children will especially enjoy meeting the Shire's tiny cousins, the Shetland ponies, as well as other small creatures at the pets corner. Any energy they have left can be worked off in the safe play area. The day can be rounded off with a visit to the coffee shop and souvenir shop. Added attractions include a breathtaking cliff top walk, nature reserve and bird sanctuary. The site has been developed with conservation in mind. A conservation plan for the farm is being worked out with the help of the North York Moors National Park.

On-Site Activities: An activity guide is given to all visitors clearly setting out the varied and interesting programme and walk.

Special Events: Fundraising activities for Save the Children fund.

Educational Facilities: Education centre; school visits.

Yorkshire Dales Falconry & Conservation Centre

Crows Nest, near Giggleswick, Settle, North Yorkshire LA2 8AS
Tel: (0729) 822832

○ Open daily 10-dusk throughout the year.
£ Adults £3.95, children £2.50, family (2 + 2) £9.50. Group rates (15 or over): adults £3.45, children £2.00.
☞ From Skipton to Settle then follow signs on A65.

Facilities: 🅿 Car and coach parking 🚻 Toilets – disabled and mother and baby facilities ☕ Refreshments 👍 Gift/book shop 🛝 Play area.

Facilities for Disabled: ♿ Wheelchair access and ramps around the centre; viewing platform for the disabled. Trails for visually handicapped. Toilets. Falconers can be assigned to assist.

Restrictions: 🐕 No dogs, but there is a special drinking well for dogs in car park.

Description: The centre has some of the most endangered birds of prey in the world. It is the only place in Britain to free-fly an Andean condor – a stunning bird with a vast three-metre wingspan, making it the largest bird of prey in the world. Other birds flown include lappet-faced vultures, golden and bateleur eagles, as well as hawks, falcons and owls from around the world. The site was purpose built, and the birds' natural habitats have been recreated as closely as possible. Children can touch some of the birds in the children's area, and see birds being fed, including new-born chicks in incubators. Visitors can learn how to fly birds at the falconry school.

On-Site Activities: Flying demonstrations four times a day; guided tours; videos in lecture room; chance to hold some of the birds.

Special Events: Conference room and falconry school.

Educational Facilities: Education officer/centre; school visits.

Special Offer: To Really Wild Guide *readers: one free child for every paying adult visiting the centre until December 1993 on production of this book.*

WEST YORKSHIRE
Lotherton Hall Bird Garden

Lotherton Estate, Townton Road, Aberford, Leeds, West Yorkshire LS25 3EB
Tel: (0532) 813723

- ○ Open daily 10-5 weekdays, 11-6 weekends.
- £ Admission free; donation box at entrance.
- ☞ A1 from Wetherby to Leeds.

Facilities: 🅿 Car and coach parking 🚻 Toilets 🍽 Refreshments 🛍 Gift/book shop ⚠ Play area.

Facilities for Disabled: ♿ Wheelchair access.

Restrictions: 🐕 No dogs.

Description: Set on a grand estate, this is one of the largest bird gardens in the country. Several paths wind through the woodland, past various ponds and lakes, to a large selection of aviaries – 100 in total – where visitors can see over 700 birds from all over the world. The first birds you are likely to spot as you enter the garden are cassowaries from Australia and New Guinea. Opposite these are a group of sarus cranes from Asia. The visitor will then wander past a large lake which is devoted to endangered breeds of waterfowl, including the very rare Ne Ne goose from Hawaii. The aviaries have all been planted with special vegetation to suit the natural habitat of each bird. There are Andean condors, black kites, flamingoes and ibis as well as several different varieties of finches, turacos and even tropical starlings.

On-Site Activities: Talks and guided tours can be arranged by staff for schools, societies and interested groups.

Special Events: International ornithological seminars held each year.

Educational Facilities: Education officer/centre; school visits.

Temple Newsam Home Farm & Estate

Temple Newsam, Leeds, West Yorkshire LS15 0AD   Tel: (0532) 645535

- ○ Farm open daily 10-4.30 in summer, 10-3.30 in winter; park and woodlands open all day throughout the year.
- £ Admission free.
- ☞ Just off A63 Leeds to Selby Road.
  Bus: No. 47 from Leeds city centre.

Facilities: 🅿 Car and coach parking   🚻 Toilets   ☕ Refreshments   🎁 Gift/book shop   🛝 Play area.

Facilities for Disabled: ♿ Limited wheelchair access. Trails for the visually handicapped.

Restrictions: 🐕 No dogs on farm.

Description: The estate is home to a Tudor-Jacobean mansion with formal gardens. This is set in 1,500 acres of parkland, farmland and woodland. Guided walks are arranged around the grounds in the summer months, and there are craft demonstrations all year round. Other attractions include a farm trail, a deserted medieval village and a Capability Brown landscape. Children's activities such as pond dipping and a woodland experience can be arranged. Historical features include an ice house, a priest's hole and Grim's dyke. There is a voluntary ranger service with opportunities for the over-fourteens to do some practical conservation work.

On-Site Activities: Nature trail; farm trail; guided walks on summer weekends; craft demonstrations.

Educational Facilities: School visits. Educational visits can be arranged with Earth Magic, pond dipping, Woodland Experience and practical work organised by the ranger. Voluntary Ranger Service with opportunities for over-fourteens to do practical conservation work.

Tropical World

Canal Gardens, Roundhay Park, Leeds, West Yorkshire LS8 1DF
Tel: (0532) 661850

- ○ Open daily 10-dusk throughout the year.
- £ Admission free.
- ☞ A58 Wetherby road, ring road.
  Bus: Nos. 10, 12, 55 from Leeds.

Facilities: ▣ Car and coach parking ♰♰ Toilets
 ☕ Refreshments 🛍 Gift/book shop.

Facilities for Disabled: ♿ Wheelchair access.
Garden for the blind.

Restrictions: 🐕 No dogs. No eating inside tropical house.

Description: A 700-acre park which is home to many species of wildlife as well as tropical animals and plants. There are several greenhouses where visitors can see a variety of tropical fish, a large collection of spiders, snakes, lizards and many rare amphibians. The butterfly house has thirty or more tropical butterflies which fly freely amongst an array of tropical plants. Soon to be completed is a nocturnal house where visitors will be able to see nocturnal animals during the day. There is a leisure lake where fishing and rowing boats are available for adults, while children can use the canoes and paddle boats. The other lake is a wildlife sanctuary for birds. Flower gardens abound with over 1,000 varieties of dahlia in one garden and more than sixty varieties of viola in another. There are several rose gardens and even a specially designed garden for the blind.

Educational Facilities: School visits. Special school pack available for teachers and worksheets for children in age groups over five, over seven and over twelve.

CHANNEL ISLANDS

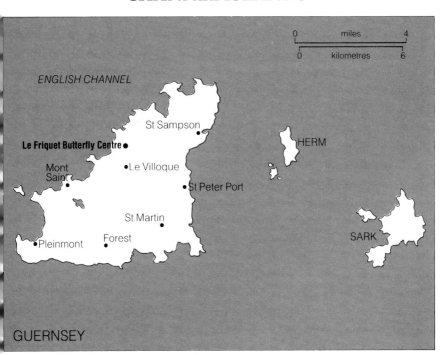

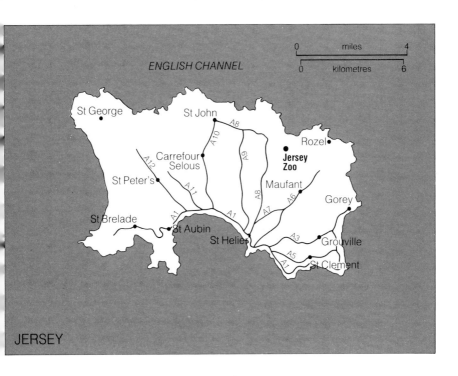

GUERNSEY
Le Friquet Butterfly Centre

Le Friquet, Castel, Guernsey   Tel: (0481) 54378

- ○ Open daily 10-5 Apr to end Oct.
- £ Adults £1.95, children £1.25, OAPs £1.50.
- ☞ Perry's Guide page 16, A3. Bus: Nos. N, F and 6.

Facilities: 🅿 Car and coach parking 🚻 Toilets ☕ Refreshments 🎁 Gift/book shop ⛰ Play area.

Facilities for Disabled: ♿ Wheelchair access.

Restrictions: 🐕 No dogs.

Description: Hundreds of exotic tropical butterflies and moths are found in the large indoor flight area. The greenhouse is filled with the food plants of these insects as well as suspended feeding bowls with banana and sugar to tempt the butterflies. Visitors can watch the caterpillars feeding, and the butterflies themselves as they hatch out of their chrysalids. Species to see include the owl, peacock, blue pansy, lacewing, giant purple emperor, clipper and orange-tip butterflies and the giant atlas moth. There is a video display within the greenhouse to explain more about these fascinating creatures. Besides butterfly-watching, visitors can also enjoy hot and cold meals in the café, mini-golf, croquet, remote-controlled toy boats and cars, boule, skittles and the children's play area.

Educational Facilities: School visits.

JERSEY
Jersey Zoo

Les Augres Manor, Trinity, Jersey JE3 5BF Tel: (0534) 864666

- ○ Open daily 10-6 or dusk throughout the year except Christmas Day.
- £ Adults £4.20, children £2.20, OAPs £2.70. Group rates: adults £3.70, children £2.00.
- ☞ North from St Helier on A7 or A8. Bus: Nos. 3a or 3b.

Facilities: 🅿 Car and coach parking 🚻 Toilets – disabled and mother and baby facilities 🍴 Refreshments 🛍 Gift/book shop 🎠 Play area.

Facilities for Disabled: ♿ Wheelchair access. Toilets.

Restrictions: 🐕 No dogs.

Description: Jersey Zoo is best-selling author and naturalist Gerald Durrell's unique sanctuary and breeding centre for endangered animals. The twenty-five acres of parkland surrounding Les Augres Manor, headquarters of the Jersey Wildlife Preservation Trust, are home to a remarkable collection of some of the world's most rare and beautiful mammals, birds and reptiles – often family groups breeding successfully. Its greatest attraction is the group of lowland gorillas, with Jambo, the head of the family, the zoo's most popular animal.

    The aim of the zoo's breeding programmes is the reintroduction of captive-bred animals back into safe places in the wild in their native homes. Jersey-born golden lion tamarinds, which inhabit a lake island at the zoo, have been reintroduced to the Brazilian jungle, and pink pigeons here have produced offspring that are now flying free in the forests of Mauritius. The lemurs of Madagascar can be seen swinging through their own woods and spectacled bears are often found sitting at the top of high poles in their enclosure.

On-Site Activities: Keeper talks and animal handling sessions; brass rubbing; face painting; activity books; touch tables; guided tours; audio-visual presentations.

Special Events: Conference groups and other special parties especially welcome. Evening visits by arrangement.

Educational Facilities: Education officer/centre; school visits.

ISLE OF MAN

Curraghs Wildlife Park

Ballaugh, Isle of Man Tel: (0624) 897323

- ○ Open daily 10-4 Easter to end Sep, Sat and Sun 10-4 in winter.
- £ Adults £2.00, children £1.00, OAPs £1.50. Group rates: 25% discount.
- ☞ On A3 halfway between Ramsey and Kirk Michael at Quarry Bends on the TT course. Bus: Nos. 5 from Ramsey, 6 from Douglas, 6a from Peel.

Facilities: 🅿 Car and coach parking 🚻 Toilets ☕ Refreshments 🛍 Gift/book shop ⛰ Play area.

Facilities for Disabled: ♿ Wheelchair access.

Restrictions: 🐕 Guide dogs only.

Description: This park is situated on peat land adjacent to the Ballaugh Curraghs, the most important wetland on the Isle of Man. It has a variety of animals and birds from Britain and overseas, but it specialises in wetland species which are on show in large walk-through enclosures, such as the South American pampas area with wildfowl, Patagonian sealions and Chilean flamingoes. Various breeds of monkeys can be seen on their island dwellings, and two species of otter. There is also a living conservation exhibit, the Ark, and a nature trail.

On-Site Activities: Guided walks and educational talks by arrangement.

Special Events: Advertised in local press.

Educational Facilities: Education centre; school visits.

NORTHERN IRELAND

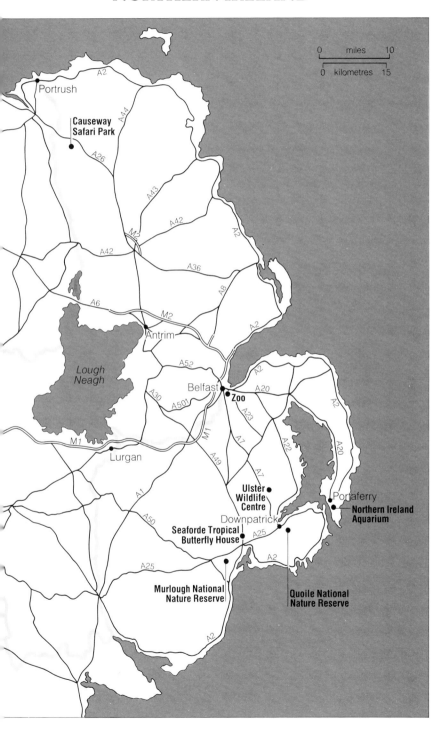

BELFAST
Belfast Zoo

Hazlewood, Antrim Road, Belfast BT36 7PN Tel: (0232) 776277

- ○ Open daily 10-6 in summer, 10-4 in winter; last admissions one hour earlier. Closed Christmas Day and closed 1 hour earlier on Fri.
- £ Adults £3.90, children £1.80, OAPs free of charge. Group rates: (25-50) 20% discount; (51-250) 25% discount; (over 250) 30% discount. Family 1-year season ticket £31.00.
- ☞ Follow signs from Carlisle Circus and Sandyknowes roundabout on M2 to 6 miles from city centre. Bus: Nos. 2, 3, 4, 6, 10, 45 from City Hall.

Facilities: 🅿 Car and coach parking 🚻 Toilets – disabled and mother and baby facilities ☕ Refreshments 🛍 Gift/book shop ⛺ Play area.

Facilities for Disabled: ♿ Wheelchair access; ramps available but site has some steep slopes. Toilets.

Restrictions: 🐕 No dogs. No alcohol. Children under 14 must be accompanied by an adult.

Description: The zoo is situated on a steep site on the face of Cave Hill, giving dramatic views out over Belfast Lough. Now reaching the end of a ten-million-pound redevelopment, the zoo is modern, conservation-conscious and progressive. Large open enclosures house colonies or family groups of animals, allowing them to live in surroundings as natural as possible. The primate house is a stimulating environment for groups of gorillas and chimps. Other attractions include a free-flight aviary, and underwater viewing of penguins and sealions. Rare animals include a family of spectacled bears, red pandas, tamarins and marmosets. Belfast Zoo has successfully bred several rare species including Hawaiian geese, Asian lions and red lechwes. Ring-tailed and black lemurs roam freely around the zoo. The education office offers illustrated talks, guided walks and handling sessions for groups of children.

On-Site Activities: Keepers are briefed to answer public questions in a friendly, polite and knowledgeable way.

Special Events: Sponsors' evenings for sponsors under the animal adoption scheme; public lectures from time to time.

Educational Facilities: Education officer/centre; school visits.

COUNTY ANTRIM
Causeway Safari Park

28 Benvarden Road, Ballymoney, County Antrim Tel: (02657) 41474

○ Open daily 10.30-6 Mar to end Sep; last admissions at 5.

£ Adults £4.00 and children £4.00. Group rates: £3.00.

☞ Just off B62 Portrush to Ballymoney road. Bus: Portrush to Ballycastle bus. Free bus from Coleraine, Portstewart and Portrush in Jul and Aug.

Facilities: 🅿 Car and coach parking 🚻 Toilets ☕ Refreshments 🎁 Gift/book shop 🛝 Play area.

Facilities for Disabled: ♿ Wheelchair access.

Restrictions: 🐕 No dogs; kennels available.

Description: This is Northern Ireland's only safari park and zoo and is home to a range of fascinating animals including lions and tigers. Visitors can take advantage of the guided coach tours through the various enclosures to gain wonderful views of the animals and learn about them from experienced guides. There is also a free indoor adventure playground and a giant slide.

On-Site Activities: Guided tours on coaches; adventure playground.

Special Events: Weekly events during July and August.

Educational Facilities: Education officer/centre; school visits.

COUNTY DOWN
Murlough National Nature Reserve

Murlough Stables, Keel Point, Dundrum, Newcastle,
County Down BT33 0NG Tel: (0396) 75467

- ○ Open access throughout the year.
  Visitor centre open daily Jun to Sep.
- £ Free admission. Car parking £1.50 Jun to Sep.
- ☞ Car park and visitor centre on A2 Dundrum to Newcastle road 2 miles north of Newcastle. Bus: Access at main car park or ¼ mile north of car park at Sliddery Ford Bridge.

Facilities: 🅿 Car and coach parking 🚻 Toilets.

Facilities for Disabled: ♿ Wheelchair access.

Restrictions: 🐕 Dogs on leads, restricted access to large parts of the reserve in summer.

Description: Set on the beautiful coast between the Mourne Mountains is a fragile 5,000-year-old sand dune system with heathland and woodland surrounded by estuary and sea. The 697-acre reserve supports a complex and delicate wildlife, rich in insects and other animals, a wide variety of plants including flowers such as the pyramidal orchid and dune burnet rose, and rare and colourful butterflies, with noteworthy populations of marsh and dark green fritillaries and wood white. The sea buckthorn of the heathland with its orange berries attracts nesting reed bunting, stonechat and whitethroat, while many species of wildfowl – including waders, ducks and swans – visit the estuary as well as some common and grey seals in late summer.

On-Site Activities: Guided walks for members of the public once a week on Wednesdays in July and August and on Sundays in September; throughout the year for visiting colleges, schools and societies.

Educational Facilities: Education centre; school visits. Visitor centre with interpretative material open June to September.

Northern Ireland Aquarium

The Ropewalk, Castle Street, Portaferry, County Down BT22 1NZ
Tel: (02477) 28062

- ○ Open Mon to Sat 10-6, Sun 1-6 Apr to Aug; Tues to Sat 10.30-5, Sun 1-5 Sep to Mar, also open bank holiday Mon
- £ Adults £1.50, children and concession 85p, family £3.85. Group rates: adults £1.25, children and concession 70p.
- ☞ 30 miles from Belfast via Newtownards and A20 (1 hour) or A2 (1½ hours) – both scenic routes. Via Saintfield A7, Downpatrick A25, Strangford and ferry boat (1 hour).

Facilities: 🅿 Car and coach parking 🚻 Toilets 🍽 Refreshments 🎁 Gift/book shop 🛝 Play area.

Facilities for Disabled: ♿ Wheelchair access.

Restrictions: 🚫 No dogs. No smoking.

Description: This is the only aquarium in Northern Ireland and it puts on show the underwater world of Strangford Lough – the largest sea inlet in the British Isles and a unique and striking area of biological importance, itself just two minutes away. In over twenty large tanks, there are displays of a changing selection of the 2,000 animals that live in the lough, from octopuses to conger eels. The touch tank is especially popular, where guides ensure visitors can stroke rays and handle sea urchins, starfish and squirting scallops. Abandoned or injured seal pups may be seen frolicking in an outside pond in spring and autumn before their release back into the lough, and in the winter months, rare loggerhead turtles, stranded by fierce winter storms, may be seen recuperating from their ordeal in a specialised tank in the aquarium.

On-Site Activities: Guides, wildfowl pond, tennis courts, putting and bowling greens, touring caravan park, adventure playground, woodland walks.

Special Events: Monthly special events for families; Conger Club for children on Wednesday afternoons in summer and Saturday mornings in winter.

Educational Facilities: Education officer; school visits. A wide range of national curriculum subjects are incorporated in general and specialised tours and education packs. Regular workshops use specialist artists and performers to make learning fun. Guided shore walks are very popular.

Quoile National Nature Reserve

5 Quay Road, Downpatrick, County Down BT30 7JB Tel: (0396) 615520

- ○ Open daily 10-5 Apr to Sep; Sat and Sun 1-5 Oct to Mar except Christmas Day, Boxing Day and New Year's Day. Other hours by prior arrangement where possible.
- £ Admission free.
- ☞ From Belfast, take A7 to Downpatrick then A25 towards Strangford and follow signs (1¼ miles). Bus: Ulsterbus No. 15 every hour from Belfast to Downpatrick bus station then No. 16 towards Strangford.

Facilities: 🅿 Car and coach parking 🚻 Toilets 🛍 Gift/book shop 🛝 Play area.

Facilities for Disabled: ♿ Wheelchair access. Braille literature will be available soon.

Restrictions: 🐕 Dogs on leads and no dogs in centre.

Description: Situated in Strangford Lough, the largest sea inlet in the British Isles, this is part of the Quoile Pondage National Nature Centre. Exhibits in the Quoile Centre show how a flood control barrage has changed the saltwater tidal estuary to a freshwater wetland with a mosaic of habitats teeming with wildlife. Vast flocks of wildfowl and wading birds converge on the Lough in winter, especially Brent geese from Arctic Canada, whooper swans from Iceland, and wigeons and teals from Eastern Europe. In summer about a third of all Ireland's terns nest in dense colonies. Riverside paths provide attractive views of the reserve and an opportunity to see the wildlife at close quarters. A bird hide is expected in 1993.

On-Site Activities: Guided walks for groups if pre-booked. Advice always available. Most groups are involved in freshwater dipping, birdwatching and general natural history walks. Equipment such as pond nets and binoculars can be borrowed.

Special Events: Programme of special events held throughout the year, generally on Sunday afternoons, on specific topics such as bats, moths, orchids etc.

Educational Facilities: Part-time education officer/centre; school visits.

Seaforde Tropical Butterfly House
Seaforde, Downpatrick, County Down Tel: (0396) 87225

○ Open Mon to Sat 10-5, Sun 2-6 Easter to end Sep.

£ Adults £2.00, children and OAPs £1.20. Group rates: children £1.00.

☞ On main Belfast to Newcastle road. Bus: from Belfast.

Facilities: 🅿 Car and coach parking 🚻 Toilets 🍽 Refreshments 🎁 Gift/book shop 🎠 Play area.

Facilities for Disabled: ♿ Wheelchair access.

Restrictions: 🐕 No dogs.

Description: Over 150 different species of free-flying exotic butterfly can be seen here, notably blue morpha, white morpha, owl, postman and zebra. The tropical setting also houses (safely behind glass) insects and reptiles from around the world including tarantulas, scorpions, iguanas, lizards, terrapins and a rock python. There is also a display panel of honey bees.

On-Site Activities: Guided tours by arrangement.

Educational Facilities: School visits.

Ulster Wildlife Centre

3 New Line, Crossgar, County Down BT30 9EP   Tel: (0396) 830282

○ Open daily 10-4 weekdays and Sun afternoons in summer. Groups by arrangement.

£ Adults £1.00, children 50p, members free of charge.

☞ Crossgar is on A7 Belfast to Downpatrick road. Centre reached by laneway beside St Columbkillie's School off Killyleagh Road. Bus: Ulsterbus Belfast to Downpatrick.

Facilities: 🅿 Car parking, coach parking by arrangement ♦♦ Toilets ☕ Refreshments occasionally 👍 Gift/book shop.

Facilities for Disabled: ♿ Wheelchair access.

Restrictions: 🐕 No dogs.

Description: At the Ulster Wildlife Centre, you can find examples of woodland, grassland, bog and wetland habitats which are home to a variety of local species. The Butterfly House is set in a Victorian walled garden and is a delightful place to visit and learn about wildlife conservation from the many displays and interpretative exhibitions. Many butterflies are attracted to the sheltered gardens and the pond hosts a huge variety of invertebrates.

On-Site Activities: Guided tours by arrangement.

Special Events: Open days and other events are publicised in the local press.

Educational Facilities: Education officer/centre; school visits.

SCOTLAND

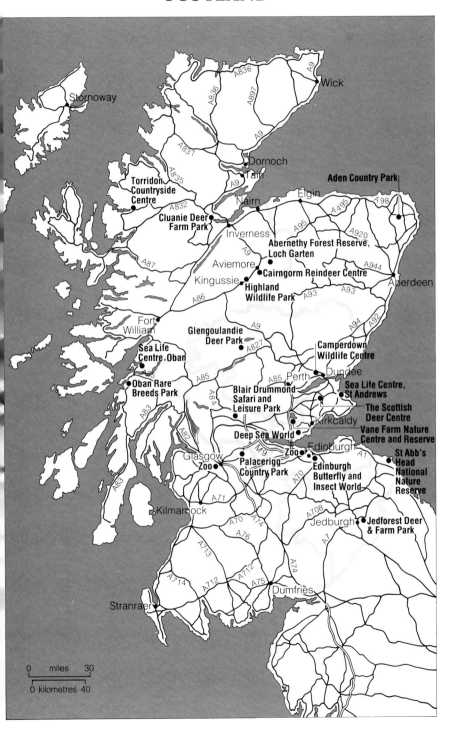

BORDERS
Jedforest Deer & Farm Park

Mervinslaw Estate, Camptown, Jedburgh TD8 6PL Tel: (08354) 364/266

- ○ Open daily 10-5.30 May to end Oct.
- £ Adults £2.20, children £1.30.
  Group rates: adults £1.95, children £1.10.
- ☞ On A68 5 miles south of Jedburgh, follow signs.
  Bus: Newcastle to Edinburgh bus on A68.

Facilities: 🅿 Car and coach parking 🚻 Toilets
🍴 Refreshments 🛍 Gift/book shop 🎠 Play area.

Facilities for Disabled: ♿ Wheelchair access.

Restrictions: 🐕 Dogs on leads.

Description: Set in 800 acres of the Jed valley with views over the Cheviot Hills, the Jedforest Deer and Farm Park is a traditional Scottish Borders working farm as well as a large conservation centre for rare breeds of sheep, cattle, pigs, goats and poultry. The farm has a flock of breeding ewes, suckler cows and red deer with some sika, muntjac and fallow. The sheep of yesteryear in the rare breeds collection include Soay from St Kilda, the most primitive breed of sheep and probably the last survivors of the prehistoric domestic sheep of Europe; the multi-horned Manx Loghtan from the Isle of Man which was recently on the verge of extinction; as well as the Hebridean, Shetland, Mouflon, South Down, grey-faced Dartmoor and Gotland. Rare breeds of cattle include the belted Galloway and the Dexter and pigs include the large black, Tamworth, Berkshire, British saddleback, Gloucester Old Spot and the middle white, Britain's rarest pig breed. In the 'clapping corner', visitors can bottle-feed lambs, stroke rabbits and calves and feed ducks and geese.

On-Site Activities: Tractor rides on request with commentary on farm, conservation and rare breeds. Guided tours available.

Special Events: National sheep-counting championship and various farm-related events throughout the year.

Educational Facilities: School visits.

CENTRAL
Blair Drummond Safari and Leisure Park

Blair Drummond, near Stirling FK9 4UR   Tel: (0786) 841456

○ Open daily 10-5.30 Apr to early Oct; last admissions at 4.30.

£ Adults £5.50, children (3-14) £3.00. Group rates (if paid one week in advance): adults £5.00, children £2.70.

☞ M9 junction 10 then follow A84 to park between Stirling (7 miles) and Callander (4 miles). Rail/Bus: Stirling station then bus to Callander and Killin or taxi to centre.

Facilities: Car and coach parking   Toilets – disabled and mother and baby facilities   Refreshments   Gift/book shop   Play area.

Facilities for Disabled: Wheelchair access; all park is on one level. Toilets. Waitress service if pre-booked. Has received good access award.

Restrictions: No dogs; kennels available.

Description: The first safari park of its kind in Scotland, and brainchild of the late Jimmy Chipperfield, the park opened in 1970. It features drive-through wild animal reserves, a boat safari round Chimps' Island, a waterfowl sanctuary, and performing sealion shows. From the elevated walkway behind the big cats' reserve, visitors can watch lions being fed. There are also Siberian tigers, zebras, camels, bison, antelopes and many other animals roaming freely. At the pets farm is a wide variety of young animals such as donkeys, ponies, piglets, calves, rabbits, wallabies and lambs.

Its original emphasis on animals in breeding groups has been expanded into other attractions where children can let off steam, such as the giant astroglide, flying-fox, adventure playground and amusement arcade.

On-Site Activities: Adventure playground; face painting; giant astroglide; pedal boat; cableslide over lake.

Educational Facilities: School visits.

FIFE
The Scottish Deer Centre
Bow of Fife, by Cupar, Fife KY15 4NQ Tel: (033 781) 391

- ○ Open daily 10-5 Apr to Oct. Groups in winter by arrangement.
- £ Adults £3.75, children £2.25, family £10.00, concession £2.75. Group rates: on application (10-15% discount).
- ☞ M90 junction 8 then follow A91 east towards St Andrews. Centre is 3 miles west of Cupar. Bus: Nos. 66 or 23 from Cupar.

Facilities: 🅿 Car and coach parking 🚻 Toilets – disabled and mother and baby facilities ☕ Refreshments 🛍 Gift/book shop 🛝 Play area.

Facilities for Facilities for Disabled: ♿ Wheelchair access. Toilets.

Restrictions: 🚫 No dogs. Kennels available.

Description: The Scottish Deer Centre is set in fifty acres of lovely parkland. At the centre is a classically designed and restored Georgian farmstead where you begin your visit at the interpretative centre. Here, an audio-visual film show and an exhibition are designed to help visitors understand all about deer – their history, life cycle and management in the wild and on the farm. There is the opportunity to hold and stroke antlers and to try to imagine what it must be like to have such things on your head. Out in the parkland, there are deer paddocks and nature walks. The canopy walks and observation platforms give an unusual aspect on the natural world; there are rangers to help answer your questions; nature trails to follow; and guided tours so that you can come face to face with the deer. Species you can hope to see are the red deer, roe deer, fallow deer, sika deer, Père David's deer, hog deer and reindeer. You can round off your day with a visit to the adventure playground.

On-Site Activities: Guided tours; nature walk; giant adventure playground.

Educational Facilities: Education officer/centre; school visits.

Sea Life Centre, St Andrews

The Scores, St Andrews, Fife KY16 9AS Tel: (0334) 74786

- ○ Open daily 10-6 Feb to Dec, 9-9 in Jul and Aug.
- £ Adults £3.95, children £2.95, concession £3.40. Group rates: 50p discount. Charges subject to seasonal changes.
- ☞ A91 (M90) 1 hour from Edinburgh, 30 minutes from Dundee, 1½ hours from Glasgow. Rail/Bus: Leuchars station then bus to St Andrews (5 miles).

Facilities: Toilets – disabled and mother and baby facilities Refreshments Gift/book shop Play area.

Facilities for Disabled: Limited wheelchair access. Toilets.

Restrictions: Guide dogs only.

Description: The centre boasts two unique features: a seal breeding colony and one of the best major displays of jellyfish in the UK. The Kriesel, a specially designed tank, produces the required rotary flow of water that keeps the jellyfish in suspension in the centre of the water mass and avoids these delicate creatures getting damaged against the hard edges of the aquarium and thus shows them in their full glory! An attraction common to all the Sea Life Centres is the sandy shoreline display. This is a large waist-high exhibit filled with many different kinds of rays that can be seen and even touched at the water surface while they flap their wings around; but beware – they are enthusiastic splashers! These rays will even take food from the fingertips of visitors.

On-Site Activities: Talks throughout the day.

Special Events: Seawatch Club is operated throughout UK Sea Life Centres. Members are entitled to free admissions to Sea Life Centres plus a newsletter and information pack. Projects are organised for Seawatch members and Small Fry Club.

Educational Facilities: Education officer/centre; school visits. Follow-up classroom visits for school groups; school competitions.

Deep-Sea World

North Queensferry, Fife, Scotland Tel: (0383) 411411

○ Open daily 9-6, except Christmas and New Year's Day.

£ Adults £4.50, children £3.00, children under 4 free, family rate (2 + 4) £13.50, OAPs Mon to Fri £3.00. Group rates and special school rates: on application.

☞ From south, follow signs for Forth Road Bridge, once across bridge take first exit and follow signs; from north off M90 at junction 1 and follow signs. Bus: regular service from Edinburgh and all parts of Scotland. Rail: North Queensferry station. Ferry: from South Queensferry.

Facilities: 🅿 Car and Coach parking 🚻 Toilets – disabled facilities ☕ Refreshments 🎁 Gift shop.

Facilities for Disabled: ♿ Disabled access throughout. Toilets.

Description: This is a new four-million-pound attraction, located beneath the Forth Rail Bridge. Its one-million-gallon tank, claimed to be the world's largest, will allow visitors a diver's eye view beneath the water by means of a 112-metre-long transparent tunnel, claimed to be the world's longest. It is to house 5,000 fish, 100 species of sharks, crustaceans and tidal fish. The largest is the porbeagle shark, which grows up to three metres, and has never before been kept in captivity. The diverse collection of sea life will also include crabs, lobsters and conger eels. There is an exhibition hall with interactive displays including shark shallows, touch tank, an octopus exhibit and a sunlit coral display.

Special Events: Facilities for conferences, meetings and special interest groups, exclusive entertaining and children's parties can be provided.

Educational Facilities: Education Officer; purpose-built classroom.

GRAMPIAN
Aden Country Park

Mintlaw, Peterhead, Aberdeenshire AB42 8FQ Tel: (0771) 22857

○ Park open 7-10 throughout the year.
Wildlife centre open weekends 2-5 May to Sep.

£ Admission free.

☞ A92 from Aberdeen. From Fochabers, take the A98 then the A950. Bus: from Aberdeen bus station.

Facilities: 🅿 Car and coach parking 🚻 Toilets ☕ Refreshments in summer 🎁 Gift/book shop in summer 🛝 Play area.

Facilities for Disabled: ♿ Wheelchair access. Garden for the visually handicapped.

Restrictions: 🐕 Dogs on leads. Two dog exercise areas available.

Description: Aden is a beautiful area of woodland, once the grounds of an estate mansion and now a country park, providing an attractive haven for people and wildlife. There is both coniferous and deciduous woodland, as well as a Victorian arboretum. The park contains a variety of plants and animals, some of which you can see on the nature trail. The trail crosses the river where you may glimpse birds such as the dipper which feed and drink around the water. The mature broadleaf forest contains some beautiful flowers such as celandines in the spring. Roe deer graze at the edges of the conifer plantation. Aden lake supports a wealth of tiny creatures like pond skaters and caddis fly larvae. Many birds, like herons, ducks and moorhens, have made their home here. The ranger service runs the wildlife centre which is open during the summer.

On-Site Activities: Nature trail and booklet from craft shop.

Special Events: Programme of special events such as guided walks and talks throughout the season.

Educational Facilities: Ranger service; school visits. Resources pack 'The Natural World of Aden with Zilla Spinner and Friends'.

HIGHLANDS
Abernethy Forest Reserve, Loch Garten

Grianan, Tullock, Nethybridge, Inverness-shire PH25 3EF Tel: (0479) 83694

- ○ Open daily 10-6 mid-Apr to mid-May, 10.30-8.30 mid-May to mid-Aug. Times depend on when ospreys arrive and settle.
- £ Adults £1.50, children 50p, OAPs £1.00.
- ☞ Follow signs from A9 at Aviemore and Carrbridge to Loch Garten. Rail: Aviemore station then Strathspey steam railway to Boat of Garten linking up with old 'classique' coach trip to reserve.

Facilities: 🅿 Car and coach parking ♣ Gift/book shop.

Facilities for Disabled: ♿ Wheelchair access.

Restrictions: 🐕 Dogs on leads in reserve. No dogs to osprey hide.

Description: Those spectacular birds of prey, ospreys, which were extinct in Scotland for many years, returned to breed here in 1959, and Loch Garten has since become one of the country's most famous and important nature reserves. There is an observation hide overlooking the ospreys, with high-powered telescopes trained on the nest. Live closed-circuit television pictures are relayed to the hide, and there are even video playbacks of recorded highlights. Young are in the nest from mid-June onwards. The surrounding land, also owned by the RSPB, includes an ancient Caledonian pine forest, interspersed with bogs, lochs and moorland, all rich in wildlife. Over fifty species of bird breed here, of which the chaffinch is the commonest, but there are also Scottish crossbills, crested tits and capercaillies. The whole area has been designated a Site of Special Scientific Interest.

On-Site Activities: Osprey centre; marked forest trails. Guided walks with warden 9.30-12.30 Wednesdays and Fridays.

Special Events: Programme of walks and talks is arranged throughout May to August.

Educational Facilities: School visits.

The Cairngorm Reindeer Centre

Reindeer House, Glenmore, Aviemore, Inverness-shire PH22 1QU
Tel: (0479) 861228

○ Open daily throughout the year. Daily visits to the reindeer at 11.

£ Adults £2.50, children £1.50.

☞ Follow Ski road east from the south end of Aviemore. Centre is 6 miles along road in Glenmore Forest Park. Bus: from Aviemore to Cairngorms, alight at Glenmore.

Facilities: 🅿 Car and coach parking 🛍 Gift/book shop.

Facilities for Disabled: ♿ Wheelchair access.

Restrictions: 🐕 No dogs on the reindeer visits.

Description: The reindeer here range freely over the northern slopes of the Cairngorm mountains. The guides take visitors amongst the herd where you can stroke and feed them. The herd breeds freely; reindeer have lived on these mountains for countless years, and they are quite at home here. All the present inhabitants were born in the UK. The reindeer feed mainly on lichens which are not grazed by other natives such as red deer. They can find their food even under snow and need no shelter, even in winter. Calves can be seen after May or June, and the reindeer are usually very tame, although the males can become aggressive during the October rut.

On-Site Activities: Guided walks, audio-visual display.

Educational Facilities: School visits.

Cluanie Deer Farm Park

Teanassie Lodge, Beauly, Highland Tel: (0463) 782415

- ○ Open daily 10-5 mid-May to mid-Oct.
- £ Adults £2.75, children £1.75, OAPs £2.25.
  Group rates: adults £2.50, children £1.50.
- ☞ On A831 Cannich road 4 miles from Beauly.

Facilities: 🅿 Car and coach parking 🚻 Toilets 🍴 Refreshments 🛍 Gift/book shop ⚠ Play area.

Facilities for Disabled: ♿ Wheelchair access. Hands-on tours available for pre-booked groups of visually handicapped visitors.

Restrictions: 🐕 No dogs.

Description: A superb natural hillside setting bordered by a river and ancient birch woodlands. Visitors to Cluanie can walk round paddocks of rare breeds of farm animals and through paddocks of red deer. Bottle-fed lambs and kids can be seen at a cuddle corner. Species on display include red, roe, fallow and muntjac deer. Wildlife likely to be seen in the area or on the nature trails includes buzzards, red kites, peregrines and many more. Visitors can watch sheep shearing, the spectacular rut of the red deer stags and the birth of baby animals in the relevant seasons. The park was nominated by *Country Living* magazine as one of the top ten open farms in Britain. There is also a small wildlife hospital specialising in deer, but local wildlife, such as hedgehogs, foxes, buzzards, owls and even a couple of pine martens have been treated here.

On-Site Activities: Environmental/nature walks; talks and special events; video show on the life of red deer. Children may be able to help feed the animals. Colour guide book available.

Educational Facilities: Education officer/centre; school visits. School groups are provided with an experienced guide; quiz sheets etc. can be tailored to tie in with school projects.

Highland Wildlife Park

Kincraig, Kingussie, Inverness-shire PH21 1NL Tel: (0540) 651270

○ Open daily 10-5 Jun to Aug, 10-4 Apr to May and Sep to Oct. For winter and evening visits, groups must pre-book.

£ Contact office for current rates.
Group rates: on application.

☞ Travelling north on A9 Inverness road, turn left at Kingussie on B9152. Travelling south on A9 Perth road, take southern Aviemore turning and follow signs on B9152. Bus: Citylink coach. Rail: Scotrail to Kingussie station (5 miles).

Facilities: 🅿 Car and coach parking 🚻 Toilets ☕ Refreshments 🎁 Gift/book shop 🛝 Play area.

Facilities for Disabled: ♿ Wheelchair access.

Restrictions: 🚫 No dogs. Kennels free of charge.

Description: The park is run by the Royal Zoological Society of Scotland and has as its theme Scottish wildlife past and present. There are two parts to the park. The main reserve is a drive-through area of about 200 acres on the south-west slope of Monadliath Mountain and features herds of red and roe deer, bison, wild horses, Soay sheep and the magnificent Highland steer. A guided trailer ride can be taken around this part of the park. The walk-around area is smaller but contains many forest and woodland habitats. In the forest are capercaillies, pine martens and viewing areas for red squirrels; while the woodland provides an ideal habitat for black grouse, tawny owls, badgers and the red fox. Other animal encounters include eagle and snowy owls, red grouse, otters, beavers, reindeer and Arctic foxes. Using its resident brown bears, the park is co-ordinating a world-wide study into improving bear enclosures.

On-Site Activities: Guided tours available for pre-booked parties.

Educational Facilities: Education officer/centre; school visits. Educational activities can be tailored to the needs of children, adults, Scouts, Guides and community groups.

Skye Serpentarium

Muilean, Harrapool, Broadford, Isle of Skye IV49 9AQ Tel: (0471) 822209

- ○ Open daily throughout the season or by arrangement. Admission usually available at any time.
- £ Adults £1.50, children 75p, family £3.50, OAPs, students and mentally handicapped £1.00. Group rates: on application.
- ☞ Follow A850 Portree road from ferry terminals at Kyleakin and Armadale (8 miles). Bus: from ferry terminals but check times with tourist office.

Facilities: 🅿 Car and coach parking 👍 Gift/book shop.

Facilities for Disabled: ♿ Wheelchair access.

Restrictions: No smoking.

Description: Housed in an old converted windmill, the serpentarium has on display a variety of reptiles and amphibians, all of which are contained in their own simulated natural habitats. Most of the reptiles are captive-bred and in the spring or summer you can witness hatchings of the incubating eggs. Snake-handling sessions are frequently held. Amongst the more exotic amphibians, brightly coloured poison-arrow frogs are kept in a forest floor habitat. Monitor lizards can be seen climbing around mangrove roots. Other animals include a number of desert lizards and tortoises. A herpetological information, education, exhibition centre and shop are part of the whole attraction.

On-Site Activities: Guided tours and handling sessions.

Educational Facilities: School visits. Emphasis is placed on the educational and conservational aspects of herpetology.

Torridon Countryside Centre

Torridon, Achnasheen, Ross-shire IV22 2EZ Tel: (0445) 791221

○ Open daily 10-5 May to Sep; deer park and deer museum open throughout the year.

£ Adults £1.00, children and concession 50p.

☞ A9 north from Inverness. Follow Ullapool signs along A835 then shortly after Garve take the A832 to Achnasheen and Kinlochewe. From Kinlochewe drive west along A896 for 10 miles to Torridon. Centre is at road junction to Torridon village (60 miles from Inverness). Rail/Bus: Kyle line train from Inverness to Achnasheen then post-bus once a day to Torridon.

Facilities: 🅿 Car and coach parking ♣ Gift/book shop.

Restrictions: 🐕 Dogs are welcome on the estate but not in the deer park or museum.

Description: The 14,000 acres of the Torridon Hills, owned by the National Trust for Scotland, can be roamed freely by the public. There is the chance of seeing some remarkable wildlife: eagles soaring, otters at play, foxes, even pine martens, are known to inhabit this area.

In the centre there is an audio-visual slide show on the scenery and wildlife of Torridon and close by is the Deer Museum which has a small but impressive collection of antlers, photographs and descriptions of the life and management of deer. Children can get quite close to the red deer herd in the centre's deer park and many can be hand-fed. It is often possible to enter the field with the hinds. The centre is also close to the seashore and a pond.

On-Site Activities: Ten-minute audio-visual slide show on Torridon, wildlife etc.

Educational Facilities: School visits.

LOTHIAN
Edinburgh Butterfly and Insect World

Melville Nursery, Lasswade, Midlothian EH18 1AZ Tel: (031) 663 4932

- ○ Open daily 10-5.30 early Mar to end Oct.
- £ Adults £2.85, children £1.60, family £8.25, concession £2.20. Group rates: about 20% discount, telephone for details.
- ☞ Follow signs on A7 Edinburgh to Dalkeith road, 300 metres south of Edinburgh bypass.
  Bus: No. 3 from shop side of Princes Street, Edinburgh.

Facilities: 🅿 Car and coach parking 🚻 Toilets – disabled facilities ☕ Refreshments 🛍 Gift/book shop 🛝 Play area.

Facilities for Disabled: ♿ Wheelchair access to entire site. Toilets.

Restrictions: 🐕 No dogs in butterfly area.

Description: The Edinburgh Butterfly and Insect World is housed in a large glasshouse which is landscaped with waterfalls, ponds and a large selection of unusual tropical plants. Butterflies from all over the world fly freely around in the area, and visitors can walk through and observe their behaviour and life cycles at first hand. The plants are carefully chosen to provide the correct caterpillar food plants so that most of the species on display are breeding at the centre. Look out for common, blue or red morons, Chinese peacocks, Palamedes, chequered or giant swallowtails, monarchs, owls, malachites, fritillaries, zebras, spotted hairstreaks and many more. Exotic insects such as leaf-cutting ants, praying mantids, stick insects, giant beetles, scorpions and tarantulas can also be seen. The whole concept of the exhibition is to show live insects behaving as naturally as possible. There is also an apiary, honey bee display and honey bee garden which is planted with a sequential arrangement of the flowering plants most important to bees in Scotland.

On-Site Activities: Guided tours for pre-booked groups of ten or more.

Educational Facilities: Education officer/centre; school visits. Guided tours for schools on topics such as butterflies and insects, the rainforest, honey bees etc.

Edinburgh Zoo

Corstorphine Road, Murrayfield, Edinburgh EH12 6TS Tel: (031) 334 9171

○   Open daily 9-6 Apr to Oct, 9-4.30 Nov to Mar. Opens 9.30 on Sun.

£   Adults £4.30, children and concession £2.30, family £12.00. Group rates (over 10): adults £3.45, children £1.85. Admission free for anyone accompanying a blind visitor.

☞   On A8 Glasgow road 3 miles west of Edinburgh city centre. Bus: Red Nos. 2, 26, 31, 69, 85 and 86 or Green Nos. 16, 18, 80, 86 and 274. Rail: Waverley or Haymarket stations then bus.

Facilities: 🅿 Car and coach parking 🚻 Toilets – mother and baby facilities ☕ Refreshments 🛍 Gift/book shop 🛝 Play area.

Facilities for Disabled: ♿ Wheelchair access. Admission free for anyone accompanying a blind visitor.

Restrictions: 🐕 No dogs.

Description: Penguins, more than any other creature, have made Edinburgh Zoo famous. In the 1950s a keeper accidentally left open a gate to the penguin enclosure and was followed around the zoo by a parade of penguins. It was to start the zoo's now famous penguin parade, an event enjoyed by thousands of visitors each year as well as the 150 or so king penguins, gentoos and macaronis which take part most afternoons. Edinburgh now has the world's largest penguin house.

The zoo is situated on a steep hill and one of the first of the zoo's 1,500 animals visitors meet as they climb are bald Waldapp ibis, Europe's rarest and most endangered bird. Further aviaries are home to an important collection of pheasants. The monkey house holds the best collection of guenons in Britain, and breeding groups of spectacled langurs, white-faced sakis and Diana monkeys. The children's playground and chimpanzee play area are opposite each other, allowing both to watch each other at play. In a rocky enclosure are polar bears. There are also brown bears and red pandas.

The reptile house has blue-tongued skinks and rainbow boas and the zoo also has many different species of bird ranging from cassowaries, parrots, flamingoes and pelicans to golden eagles. Cats include margays, lions, tigers, leopards and jaguars. Right at the top of the hill, with views across the city, is the African plains exhibit with herds of scimitar-horned oryx, red lechwe and zebra.

On-Site Activities: Daily talks at gorilla, chimpanzee, giraffe, penguin, lion, rhino and sealion enclosures with emphasis on conservation. Animal handling classes every weekend from Easter to end August and throughout the Lothian school holidays. Evening guided tours can be booked in advance for clubs, groups and societies from early May to end August. Penguin parade outside penguin enclosure daily from April to September (or later, weather permitting).

Educational Facilities: Education officer/centre; school visits.

STRATHCLYDE
Glasgow Zoo Park

Calderpark, Uddingston, Glasgow G71 7RZ Tel: (041) 771 1185

○ Open daily 10-6 in summer, 10-5 in winter.

£ Adults £3.45, children (3+), OAPs, students, mentally handicapped adults, unemployed £2.10, mentally handicapped juveniles £1.00, carers with mentally handicapped £2.60, family (2 + 2) £9.00. Group rates (16 or over): adults £2.60, children (2-16) £1.55.

☞ Off ramp at junction 4 on M74 to Glasgow south-east, 5 miles from Glasgow city centre at end of London Road. Bus: Anderston bus station (platform 14) Nos. 43, 44, 240; (platform 16) No. 250.

Facilities: 🅿 Car and coach parking 🚻 Toilets – mother and baby facilities ☕ Refreshments 🎁 Gift shop 🛝 Play area.

Facilities for Disabled: ♿ Wheelchair access.

Restrictions: 🐕 Guide dogs only.

Description: The zoo is home to a multitude of animals and boasts the best bear enclosure in the world, housing the endangered Asian black bear. There is a large collection of cats including lions, tigers, cheetahs, margays and ocelots. White rhinos, Père David's deer, camels, llamas and porcupines can also be discovered. Monkeys include lemurs, capuchins, cotton-topped tamarins and Sulawesi crested macaques. The tropical house contains a magnificent collection of reptiles ranging from Carolina Box tortoises to Cuban pythons. There have been a number of successful breeding programmes and enormous effort has been put into cage design. Other features include an eagle owl aviary, an education centre, children's playground and souvenir shop.

On-Site Activities: Excellent conservation-orientated guide book.

Educational Facilities: Education officer/centre; school visits. Education service provides guided tours, talks, animal handling, close contact with animals etc.

Oban Rare Breeds Park

New Barran, Oban, Argyll PA34 4QD Tel: (063177) 604/8

- ○ Open daily 10-5.30 Mar to mid-Jul and Sep to end Oct; 10-7.30 mid-Jul to end Aug.
- £ Adults £2.50, children £1.50, family £7.00, OAPs £1.50. Group rates: 10% discount.
- ☞ On Glencriutten road 2 miles from Oban or 3 miles from Connel or 3½ miles from Kilmore; follow signs.

Facilities: 🅿 Car and coach parking 🚻 Toilets ☕ Refreshments 🛍 Gift/book shop.

Facilities for Disabled: ♿ Wheelchair access to some areas.

Description: A thirty-acre park set in the beautiful scenery of the west Highland countryside. Wander along the woodland walks and then out on to the hills to gain spectacular views of Glencoe. There is a large collection of rare breeds of farm animals – sheep, poultry, pigs and cattle, as well as a small herd of tame red deer. Children can delight in feeding the farm animals themselves and then head off to the pets corner where there is a selection of friendly and cuddly baby deer, goats, rabbits and lambs. End your visit by relaxing in the tea room which boasts excellent home baking.

Educational Facilities: School visits.

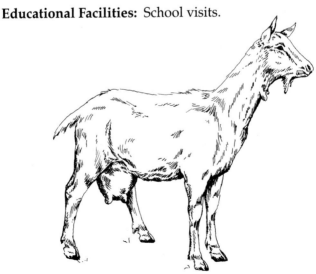

Palacerigg Country Park

Cumbernauld G67 3HU Tel: (0236) 720047

○ Open daily 10-6 Apr to Sep, 10-4.30 Oct to Mar; closed Mon and Tues in winter.

£ Admission free.

☞ Follow Cumbernauld sign off A80 Glasgow to Stirling road then Palacerigg signs.

Facilities: Car and coach parking Toilets Refreshments Gift/book shop Play area.

Facilities for Disabled: Wheelchair access.

Restrictions: No dogs, guns or fires.

Description: The theme of Palacerigg Country Park, which covers 750 acres of woodland, farmland and moor, is Scottish and European animals. The park also has an impressive range of rare and minority breeds, particularly sheep varieties – Shetland, North Ronaldsay, Hebridean, Soay, Jacob and Lewis blackface – and poultry – black East Indian duck, magpie duck, cayuga duck, harlequin duck, Indian runner duck, Shetland goose, Polish bantam chicken, Scots dumpy chicken and Old English game chicken. Badgers, stoats, weasels and sometimes mink are also found around the park. The only true reptilian resident at Palacerigg is the common lizard, but amphibians include the common frog, toad, smooth newt, palmate newt and great crested newt. Of the four species of deer at the park, only the roe lives wild. There are wild nesting long-eared owls as well as little owls, snowy owls and eagle owls in large enclosures and an observation hide overlooking Fannyside Loch.

On-Site Activities: Guided walks; farm demonstrations.

Special Events: Easter animal parade; sheep-shearing event; teddy bears' picnic.

Educational Facilities: Education officer/centre; school visits.

St Abb's Head National Nature Reserve

St Abb's, Berwickshire TD14 5QF Tel: (08907) 71443

- ○ Reserve open throughout the year. Centre open daily 10-5 Apr to Sep.
- £ Admission free.
- ☞ From A1 take A1107 Eyemouth coast road to Coldingham then B6438 to St Abb's. Bus: taxi-bus from Berwick to St Abb's via Eyemouth and Coldingham.

Facilities: 🅿 Car and coach parking ♦♦ Toilets ☕ Refreshments 🛝 Play area.

Facilities for Disabled: ♿ Limited wheelchair access.

Restrictions: 🐕 Dogs on leads. Groups must pre-book.

Description: St Abb's Head is the largest seabird colony in south-east Scotland with 80,000 cliff-nesting birds including guillemots, razorbills, fulmars, shags, kittiwakes and small numbers of puffins, present between April and July. It is also an excellent site for seeing migrant birds both on land and at sea – warblers, flycatchers, chats, shearwaters and skuas. But this 192-acre National Nature Reserve is also important for its geology and its plant and invertebrate life. The reserve, which supports over 250 species of flowering plant, has a thriving population of the rare Scots lovage and the uncommon purple milk-vetch. The sea off St Abb's Head is very rich in marine animals and plants, with an unusual mix of species of both Atlantic and Arctic origins. Cold water animals like the wolf fish and the lovely bolocera anemone are found here at the limits of their southern range.

On-Site Activities: Ranger-led walks; boat trips.

Educational Facilities: School visits.

Sea Life Centre, Oban

Barcaldine, Oban, Argyll PA37 1SE Tel: (0631) 72386

- ○ Open daily 9-6 Feb to Nov, 9-7 Jul and Aug, weekends Dec and Jan.
- £ Adults £3.95, children £2.95. Group rates: on application.
- ☞ On A828 Oban to Fort William road, 10 miles north of Oban. Bus: from Oban.

Facilities: P Car and coach parking ♦♦ Toilets – mother and baby facilities ⚑ Refreshments ♣ Gift/book shop ⛫ Play area.

Facilities for Disabled: ♿ Wheelchair access with some assistance.

Restrictions: ✘ No dogs.

Description: The Oban Sea Life Centre offers visitors a chance to meet one of the most appealing of our British marine mammals, the common seal. While enjoying the antics of the seals at close hand there is also the opportunity to learn of the rear and release programme for baby seals that the centre operates. Abandoned and stranded common seal pups are cared for here and, when the time is right, released back into the wild. Another unique feature at Oban is the herring ring, a giant doughnut-shaped aquarium display in which visitors view shoals of herring and other fish in a never-ending circuit. Steps take the onlooker right underneath the display and up through the very middle of a swirling mass of shimmering fish.

On-Site Activities: Twelve talks/feeds each day on various marine topics.

Educational Facilities: Education officer/centre; school visits.

TAYSIDE
Camperdown Wildlife Centre

Camperdown Country Park, Dundee DD2 4TF Tel: (0382) 623555

- ○ Open daily 10-4.30 throughout the year; last admissions at 3.45. Extended hours in public holidays.
- £ Adults £1.10, children and OAPs 80p. Group rates: 60p.
- ☞ In north-west corner of Dundee, entrance off A923 Coupar Angus road. Bus: from city centre.

Facilities: Car and coach parking Toilets – mother and baby facilities Refreshments Gift/book shop Play area.

Facilities for Disabled: Wheelchair access.

Restrictions: No dogs. Children must be accompanied by an adult.

Description: Surrounded by beautiful mature woodland, this centre contains one of the finest collections of Scottish and European wildlife in the country. Scottish fauna is its speciality and this includes residents of the past – the brown bear, timber wolf, Arctic fox and reindeer – as well as those of the present – deer, golden eagle, Scottish wildcat, snowy owl and Britain's rarest mammal, the pine marten. Some have recently made Scotland their home, such as the Manchurian sika deer, the muntjac deer, the mink, even the fallow deer, once also a foreign species but now seen wild at Camperdown. There is also an impressive collection of rare breeds. Sheep include the primitive Soay; goats range from the little pygmy to the golden Guernsey; and there are many ornate birds. One of the strangest is the turkey-like capercaillie, once native to Scotland, reintroduced and struggling to survive. Four were bred at the centre in 1991.

On-Site Activities: Guided tours for groups; animal handling days.

Special Events: Creepy-crawly days; falconry displays.

Educational Facilities: School visits. Covered area for school groups available in May and June.

Glengoulandie Deer Park

Glengoulandie, Foss, by Pitlochry, Perthshire PH16 5NL
Tel: (0887) 830306/509

○ Open daily 9-9 May to Sep, 9-6 in winter.

£ Adults 95p, children 65p, cars £3.00.
Group rates: on application.

☞ On B846 from Aberfeldy to Kinlock Rannoch, 8 miles north of Aberfeldy.

Facilities: P Car and coach parking ♦♦ Toilets
🛍 Gift/book shop ⚠ Play area.

Restrictions: 🐕 Dogs on leads. Visitors stay in their cars.

Description: Glengoulandie Deer Park lies at the foot of Schiehallion, the 'magic mountain' in highland Perthshire, and features a fine herd of red deer, including one of the finest stags in Scotland. A scenic roadway enables visitors to drive through the park, which also serves as a natural reserve for preserving the blood lines of some rare breeds of sheep. There are also Highland cattle, tame and wild geese, mallard and muscovy ducks and friendly goats.

Educational Activities: School visits.

Vane Farm Nature Centre and Reserve

by Loch Leven, Kinross KY13 7LX Tel: (0577) 862355

- ○ Open daily 10-5 Apr to Dec, 10-4 Jan to Mar.
- £ Adults £1.50, children 50p, concession £1.00. RSPB and YOC members free of charge.
- ☞ On B9097 1½ miles east of M90 junction 5 Edinburgh to Perth road.

Facilities: 🅿 Car and coach parking 🚻 Toilets
☕ Refreshments in Jul and Aug and weekends
🛍 Gift/book shop.

Facilities for Disabled: ♿ Wheelchair access. Trails for visually handicapped.

Restrictions: 🐕 No dogs.

Description: From this nature centre and reserve, there are fine views over Loch Leven. Some of the most spectacular sights on the loch can be seen between October and December when some 15,000 pink-footed geese arrive from their Icelandic breeding grounds to winter here. Before Christmas more than 200 whooper swans can also be seen. Early signs of spring on the loch are the arrival of large flocks of curlews and the reappearance of shelducks, gudwalls, oystercatchers and redshanks. In the autumn all five species of grebe have been seen at Vane Farm.

On the 298-acre reserve itself, a nature trail climbs to the top of Vane Hill. In the summer the woods hold over thirty pairs of willow warblers, five to six pairs of tree pipits and several redpolls. There are good numbers of green-veined white butterflies and in the autumn some spectacular fungi such as the red-and-white fly agaric.

On-Site Activities: Guided walks on Mondays in July and August.

Special Events: Special events during the summer, most with some aspects aimed at children.

Educational Facilities: Education officer/centre; school visits. School visits are taken by a teacher naturalist and are designed to fit in with curricular activities and subjects.

WALES

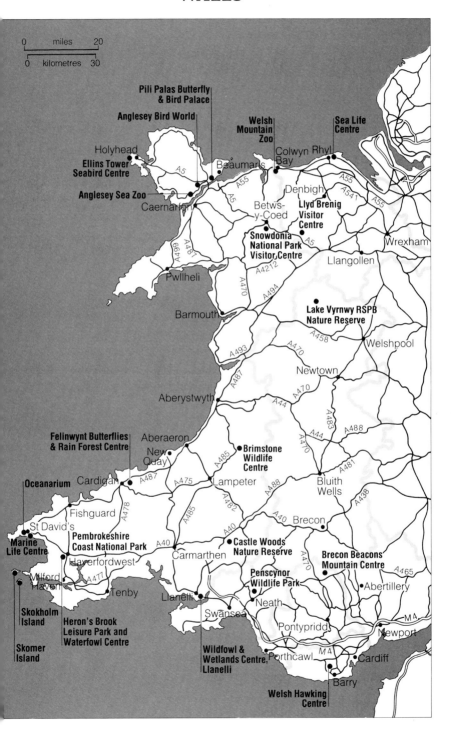

CLWYD
Llyn Brenig Visitor Centre

Cerrigydrudion, Corwen, Clwyd LL21 9TT Tel: (049 082) 463

○ Open daily 9-dusk Apr to end Oct; Mon to Fri 8-4 Nov to Mar.

£ Admission free. Car parking 50p

☞ On B4501 between Denbigh and Cerrigydrudion.

Facilities: 🅿 Car and coach parking 🚻 Toilets – disabled facilities ☕ Refreshments 🛍 Gift/book shop ⛱ Play area.

Facilities for Disabled: ♿ Wheelchair access. Toilets. Fishing boats for the disabled.

Description: This is a reservoir lying in a tranquil Welsh valley in the heart of the Denbigh Moors. Since the completion of the Brenig dam in 1976 it has provided a valuable resource for the angler, the naturalist and the sportsperson. The wide range of activities on offer include hides, giving the birdwatcher views of wildfowl in autumn. Leaflets are also available for nature trails exploring the streams, reservoir, woodland and moorland landscape and archaeological trails which delve into the secrets of Bronze Age burial rites. The moorland, forest and lake habitat means that many birds, such as redpoles, crossbills, goosanders, buzzards and even ospreys can be seen flying on migration over the area in March and April.

On-Site Activities: Guided walks; nature trail; orbital exhibition and audio-visual display.

Educational Facilities: School visits.

Sea Life Centre, Rhyl

Rhyl, Clwyd LL18 3AP   Tel: (0745) 344660

○ Open daily 10-6 throughout the year; last admissions at 5.

£ Adults £3.85, children £2.75, OAPs £2.95. Group rates: on application.

☞ Follow signs in Rhyl.

Facilities: 🅿 Car and coach parking 🚻 Toilets – disabled and mother and baby facilities ☕ Refreshments 🛍 Gift/book shop.

Facilities for Disabled: ♿ Wheelchair access. Toilets.

Restrictions: 🐕 No large dogs.

Description: Under the banner, A Unique Deep Sea Adventure, the Sea Life Centre at Rhyl surrounds the visitor with hundreds of gallons of water, in tanks of course, full of examples of British sea life – from octopuses and conger eels to starfish, sharks and a whole variety of rays. Each tank has a particular theme, from The Living Seas to The Impact of Man. The highlight of the centre must be its underwater tunnel allowing visitors an uninterrupted view of life at the bottom of the sea, confronting them with an astonishing array of curious creatures. In addition, there are usually displays and exhibitions on elements of sea life from an educational angle. The centre is completely covered and so it is comfortable to visit all year round.

On-Site Activities: Feeding displays; touch-pool talks; regular talks and demonstrations.

Educational Facilities: Education officer/centre; school visits. Annual teachers' evenings. Special project packs are available which are tailored to the national curriculum.

Welsh Mountain Zoo

Old Highway, Colwyn Bay, Clwyd LL28 5UY Tel: (0492) 532938

- ○ Open daily 9.30-5 mid-Mar to Oct, 9.30-4 Nov to mid-Mar.
- £ Adults £4.40, children and OAPs £2.20, wheelchair-bound and blind visitors free. Group rates: adults £3.30, children and OAPs £1.65, 1 adult free with every 10 full-paying children.
- ☞ Leave Colwyn Bay's West End shopping area on King's Road B5113. The zoo is signposted from the A55 Colwyn Bay bypass at the Rhos-on-Sea interchange. Train/Bus: Free mini-bus from Colwyn Bay station Apr to Sep.

Facilities: 🅿 Car and coach parking 🚻 Toilets – disabled and mother and baby facilities 🍽 Restaurant, bistro bar and picnic sites 🛍 Gift/book shop ⛱ Play area.

Facilities for Disabled: ♿ Wheelchair access to 70 per cent of zoo. Free admission to wheelchair-bound and blind visitors. Toilets.

Restrictions: 🐕 Guide dogs only.

Description: The Welsh Mountain Zoo houses many animals in natural surroundings. There are examples of rare British wildlife such as otters, red squirrels, polecats and pine martens. There are many exotics like elephants, lions, bears, deer, chimpanzees and monkeys, rare Persian leopards, ostriches, flamingoes, sealions, penguins and tropical birds. Youngsters will love the children's farm, as well as exploring jungle adventureland, alligator beach, the reptile house and woodland nature trails. Everyone can relax in the restaurant or licensed bistro. The chimp encounter is a multi-media display which shows you all about chimps and their threatened habitat. The Californian sealions continue to delight audiences with their energetic displays. The emphasis is on conservation and education put across in a fun and entertaining fashion.

On-Site Activities: Woodland nature trails. Guided tours available in autumn, winter and spring. Special talks and themed days.

Educational Facilities: Forty-seater education centre with stage, projector and tape facilities. Touch and learn corner. Full-time education officer. Themed worksheets for different age groups. All education facilities geared to a cross-curricular theme.

DYFED
Brimstone Wildlife Centre

Penuwch, near Tregaron, Dyfed SY25 6RA Tel: (097 423) 439

- Open daily 10-6 Easter to end Oct. Pony trekking and bistro open throughout the year.
- £ Adults £2.50, children and disabled £1.50, OAPs £2.00. Pony trekking £6.00 per hour.
  Group rates: on application.
- Off A487 Aberystwyth to Aberaevon coast road at Llanrhystud and follow B4337 to Cross Inn. Turn right on B4577 through Penuwch.

Facilities: Car and coach parking Toilets Refreshments Gift/book shop Play area.

Facilities for Disabled: Wheelchair access.

Restrictions: Guide dogs only.

Description: The centre features a large tropical conservatory, home to about seventy-two species of butterfly. Visitors can wander through this simulated rainforest and observe parrot and zebra finches free-flying alongside the butterflies. Waterfalls flow into ponds inhabited by koi carp, goldfish and terrapins. Tarantulas are also on display.

  A separate aviary holds five species of bird including cockatiels and love birds. Outside, a collection of rare breeds can be discovered including pot-bellied pigs and sheep. Favourites amongst the children are the rabbits, chipmunks and guinea pigs. An educational horse show takes place in the arena: different breeds from miniature Shetland ponies up to Shire horses are paraded whilst the commentator explains the significance of each. Pony trekking and rides are available. Lakes and wetlands provide ideal habitats for various wild flowers and waterfowl.

Special Events: Full and half day activities available for children including helping with pets, pony riding and picnics.

Educational Facilities: School visits.

Castle Woods Nature Reserve

Llandeilo, Dyfed Tel: (0558) 822370

- ○ Open daily. Information centre open weekends and bank holidays in summer.
- £ Admission free.
- ☞ From car park next to fire station, follow Church Street to river. Entrance at bridge – follow badger footprint signs. Car park on A40. Bus: from Carmarthen or Llandovery to Llandeilo.

Facilities: 🅿 Car parking 👍 Gift/book shop.

Restrictions: 🐕 Dogs on leads.

Description: The reserve lies west of Llandeilo and comprises two areas of woodland along a steep escarpment overlooking the River Tywi. To accompany the five quarries on the estate, the ruins of Dinefwr Castle stand proudly on a hill rising from the Tywi water meadows. In the spring the ground is carpeted with wood anemones, bluebells, dog's mercury and primroses as well as being rich in lichens, including the large lungwort. Such a vast quantity of mature woodland provides the ideal habitat for great spotted, lesser spotted and green woodpeckers. Nuthatches and treecreepers also forage amongst the bark for food. During the summer it is possible to see redstarts, pied and spotted flycatchers. The water meadows below the woods provide a home for wintering wildfowl such as mallard, teal, wigeon, goosander, shoveller and pochard. There are several active badger sets and foxes, hares, rabbits, fallow deer, grey squirrels and several species of bat are also permanent residents.

On-Site Activities: Guided walks.

Educational Facilities: Education officer/centre; school visits.

Felinwynt Butterflies & Rain Forest Centre

Rhosmaen, Felinwynt, Cardigan, Dyfed SA43 1RT Tel: (0239) 810882

○ Open daily 10.30-5 end May to end Sep.

£ Adults £1.75, children 75p, OAPs £1.50.
Group rates: on application.

☞ From A487 5 miles north of Cardigan follow signs to Felinwynt.

Facilities: 🅿 Car and coach parking 🚻 Toilets 🍽 Refreshments 🛍 Gift/book shop.

Facilities for Disabled: ♿ Wheelchair access.

Restrictions: 🐕 Dogs welcome on site but not in butterfly house.

Description: A small butterfly centre designed to represent a tropical rainforest, this is complete with recorded sound effects to enhance the atmosphere. During the season, it houses twenty to thirty species, mainly from south-east Asia and South America, including a colony of zebra, owl with its eighteen-centimetre wingspan, and scarlet swallowtail. Many breed there successfully, so there are plenty of caterpillars and pupae, showing visitors the entire life cycle. There are also common frogs and toads, several species of stick insect, and Chinese painted quail which help control pests and add to the interest. A small exhibition describes rainforests and the threat of destruction they face.

On-Site Activities: Guided tours.

Educational Facilities: School visits.

Heron's Brook Leisure Park and Waterfowl Centre

Heron's Brook, Bridge Hill, Narberth, Dyfed SA67 8QX Tel: (0834) 860723

- ○ Open daily 10-6 Easter to end Sep.
- £ Adults £2.00, children £1.25, OAPs £1.50. Group rates: 15% discount.
- ☞ Follow A40 from Carmarthen then take A478 to Narberth and follow signs.
  Rail: Carmarthen to Narberth station (1 mile).

Facilities: 🅿 Car and coach parking 🚻 Toilets – disabled and mother and baby facilities ☕ Refreshments 🎁 Gift/book shop 🎠 Play area.

Facilities for Disabled: ♿ Wheelchair access. Toilets.

Restrictions: 🐕 Dogs on leads.

Description: The centre has thirty acres of picturesque parkland and deciduous woodland with over forty breeds of domestic and ornamental waterfowl and pheasant. During the breeding season there is the opportunity to visit the rearing house where techniques including incubation hatching can be observed; Heron's Brook breeds in excess of 500 birds each year. A woodland walk leads to the cleverly designed Brock's Hide enabling internal and external viewing of a badger set. Pets corner allows the visitor to get involved with feeding some of the animals, and some of the most endearing at that! Twice a day you can bottle-feed baby calves, molly lambs, pygmy kids and even pot-bellied pigs. At Bunnyland, the guinea pigs and rabbits can also be fed. Eighty per cent of all the animals and birds at Heron's Brook are free-roaming, which without doubt adds to the enjoyment of the centre.

On-Site Activities: Woodland walk, pony rides, milking demonstration, supervised animal feeding times.

Special Events: Advertised locally.

Educational Facilities: School visits; school education pack.

Marine Life Centre

Feidr Pant-y-Bryn, St David's, near Haverfordwest, Dyfed SA62 6QS
Tel: (0437) 721665

○ Open daily 10-5 Mar to end Oct.
  Open for group bookings by appointment out of season.

£ Adults £2.50, children and OAPs £1.50, family £8.00.
  Group rates: 10% discount.

☞ From Haverfordwest follow A487 and centre is on left-hand side as you enter St David's. From Fishguard follow A487 through St David's on the Haverfordwest road; the centre is on the right-hand side as you leave St David's. Bus: from Haverfordwest or Fishguard to St David's daily except Sun and public holidays.

Facilities: 🅿 Car and coach parking 🚻 Toilets – mother and baby facilities 🍴 Refreshments 🛍 Gift/book shop 🛝 Play area.

Facilities for Disabled: ♿ Wheelchair access.

Restrictions: 🐕 Dogs on leads if kept under control. No dogs in cafeteria.

Description: At the Marine Life Centre, you can imagine yourself beneath the sea as you wind your way through the aquarium of local marine creatures in their carefully prepared environment which simulates underwater caves. The inhabitants include octopuses, sea scorpions, lobsters, suckerfish, conger eels and pipefish among others, plus members of the shark family such as dogfish, huss and rays. You can enjoy a diver's eye view at the 5,000-gallon shipwreck tank where fish swim among the broken spars and debris of an old vessel. There are Use of the Sea and conservation displays, video presentations on marine life, a large wave tank and an ex-RNLI lifeboat within the grounds.

On-Site Activities: Staff are on hand to guide people through the aquarium and answer their questions.

Educational Facilities: Education officer/centre; school visits.

Oceanarium

42 New Street, St David's, Pembrokeshire SA62 6SS Tel: (0437) 720453

○ Open daily 10-6 in summer, 10.30-4 in autumn, winter and spring.

£ Adults £2.25, children £1.50, family (2 + 2) £6.50, family (2 + 3) £7.00. Group rates (over 15): adults £1.75, children £1.20.

☞ Follow A487 to St David's then follow signs, 100 metres from city centre. Rail/bus: Haverfordwest or Fishguard stations then bus to the Oceanarium in St David's.

Facilities: P Car and coach parking ♚ Toilets ☕ Refreshments ♣ Gift/book shop ⚒ Play area.

Facilities for Disabled: ♿ Wheelchair access to 60 per cent.

Restrictions: 🐕 Dogs on leads.

Description: A closed-system aquarium, the Oceanarium uses no chemicals, artificial sea water or filters but maintains a constantly recirculated 250,500 gallons (360,000 litres) of sea water and does not pump directly from the sea. The London Zoological Society has described the centre as unique in its aim to sustain the natural life cycles of the marine life it exhibits. Housed on two floors, the large-scale aquarium includes a panoramic tank beautifully displaying all manner of marine life, a shark and ray tank with upstairs viewing galleries and a six-metre touch tank giving the visitor a chance to come into close contact with a variety of sea creatures. As well as supervising the sessions at the touch tank, expert aquarists are on hand to answer questions and provide a further insight into what may be seen while touring the centre.

On-Site Activities: Full talk during main school holidays. Other times subject to visitor numbers.

Educational Facilities: School visits.

Pembrokeshire Coast National Park

County Offices, St Thomas Green, Haverfordwest, Dyfed SA61 1QZ
Tel: (0437) 764591

Description: Here, at the most western extremity of Wales you can find rugged cliffs, broad bays and islands, along with tree-lined creeks in the Daugleddi area, and open moorland on the Preseli Mountains. Thanks to the mild climate, wild flowers flourish. The 180-mile coast path provides superb views of wild rocks and sandy beaches, and there are nature reserves on the islands and moors as well as renowned seal and seabird colonies.

In its 225 square miles (580 sq kms) you will also find farms and woodlands. The park is home to a whole range of wildlife, much of it still surviving as it did in the Stone Age.

The park has seven information centres in the area, including one in Newport and another in St David's. Each year some 300 events are organised to improve public awareness of the local environment.

Skokholm Island

c/o Dyfed Wildlife Trust, 7 Market Street, Haverfordwest, Dyfed SA61 1NF
Tel: (0437) 765462

- ○ Open Apr to end Sep. For day visitors mid-Jun to mid-Aug.
- £ £5.00.
- ☞ Follow Dale road from Haverfordwest for 10 miles, cross marshy area then turn right through Marloes to National Trust car park and walk (5 minutes) to boat.

Facilities: Car and coach parking　Toilets　Gift/book shop.

Restrictions: No dogs.

Description: Skokholm Island is a naturalists' holiday destination, an area of Special Scientific Interest, open to anyone interested. Stays are usually for not less than one week, full board is provided and there are several study courses each season. The flora of the island is that usually found in sub-maritime grassland, but the population of birds is special and famed for its communities of Manx shearwaters and storm petrels and for its strong colonies of puffins, razorbills and guillemots. The island is also the breeding station for up to sixty pairs of oystercatchers, but is particularly well known for its migrants such as chiff chaffs, willow warblers, whitethroats, spotted and pied flycatchers. Some rare species which have been recorded include spoonbills, hobbys, great skuas, alpine swifts and hoopoes, to name a few.

On-Site Activities: Guided walks for day visitors. For those staying a week, courses in art, photography, migrant birds and island ecology are arranged.

Skomer Island

c/o Dyfed Wildlife Trust, 7 Market Street, Haverfordwest, Dyfed SA61 1NF
Tel: (0437) 765462

○ Open daily Apr to end Oct except Mon. Open bank holidays.

£ £5.00.

☞ Follow Dale road from Haverfordwest for 10 miles, cross marshy area then turn right through Marloes to National Trust car park and walk (5 minutes) to boat.

Facilities: Car and coach parking Toilets.

Facilities for Disabled: Wheelchair access.

Restrictions: No dogs.

Description: There are 720 acres of wild and spectacular scenery on Skomer, legendary to all birdwatchers and possibly the most important seabird site in the south of Britain. It is only when approaching the island that the visitor may realise how many birds inhabit such a small area. On landing it is usually possible to see puffins, razorbills, guillemots, kittiwakes and numerous gulls. The Manx shearwater colony is perhaps the largest of this species in the world. Large numbers of other scarce mainland species can be found including chough, curlew, lapwing, raven, rock pipit and sedge warbler. The best time of the year to see the seabird colonies is April to mid-August, and migrant birds can be seen in April/May and August/September. Flowers are at their best in May, June and July. The unique Skomer vole is not to be missed and seals can be seen throughout the year in Pigstone Bay. Footpaths lead to all the best viewing sites.

On-Site Activities: The warden or assistant warden meets all visitors to give a brief talk on the island before they are allowed to wander at their own pace.

Educational Facilities: School visits.

The Wildfowl & Wetlands Centre, Llanelli

Penclacwydd, Llwynhendy, Llanelli, Dyfed SA14 9SH
Tel: (0554) 741087

- ○ Open daily 9.30-5 in summer, 9.30-4 in winter except Christmas Eve and Christmas Day.
- £ Adults £2.95, children £1.50, family £7.50.
  Group rates: about 20% discount. WWT members free.
- ☞ 3 miles east of Llanelli, south of A484 Llanelli to Swansea road, 1 mile west of Loughor Bridge.
  Bus: from Llanelli or Swansea to Llwynmendy (1 mile).

Facilities: 🅿 Car and coach parking 🚻 Toilets – parent and baby facilities ☕ Refreshments 🎁 Gift/book shop 🎠 Play area (1993).

Facilities for Disabled: ♿ Wheelchair access. Hide access for the disabled.

Restrictions: 🐕 No dogs.

Description: Special hides and level walkways at this fascinating wildlife centre provide spectacular views of a host of wild birds, with over 1,000 captive waterfowl from around the world. Here you can find Hawaiian geese, bright pink Caribbean flamingoes and rare white-winged wood ducks. Flocks of curlews, oystercatchers and redshanks can be seen wheeling over the 150-acre saltmarsh forming part of this Site of Special Scientific Interest. There are wintering pintail, wigeon and teal keeping watch for the peregrine falcon. At the summer duckery you can hear ducklings calling from inside their eggs.

On-Site Activities: Birdwatching tours; hatchery tours in summer; wetland plant walks; high tides etc.

Educational Facilities: Education officer/centre; school visits. Education programmes are linked to school work and can be tailor-made.

GWYNEDD
Anglesey Bird World

Tynparc, Dwyran, Llanfairpwll, Gwynedd Tel: (024 879) 627

- ○ Open daily 10-6 Easter to end Sep.
- £ Adults £2.75, children and OAPs £1.75, family (2 + 3) £8.00. Group rates: on application.
- ☞ On A4080.

Facilities: Car and coach parking Toilets – disabled and mother and baby facilities Refreshments Gift/book shop Play area.

Facilities for Disabled: Wheelchair access. Toilets.

Restrictions: No dogs.

Description: The Bird World and children's pleasure park has a small collection of paddocked animals – goats, sheep, donkeys and pigs – and a duck pond. It also houses an exotic collection of birds from four continents and a group of very friendly falcons. Other residents are Australian black swans, Indian runners and North American wood ducks. The walk-in aviaries allow visitors to mingle with the birds. The wildfowl are exceptionally tame and can be hand-fed. The eight-acre site incorporates a toddlers' play area, free bouncy castle, a large selection of toys, including mini-trikes, scooters, rocking boats, barrel slides and see-saws, all contained in a safe play area within sight of parents. There are also donkey rides and a ride to Swan Lake aboard the Swan Lake miniature railway.

On-Site Activities: Donkey rides; railway rides; guided tours on request.

Educational Facilities: School visits.

Anglesey Sea Zoo

Brynsiencyn, Anglesey, Gwynedd LL61 6TQ Tel: (0248) 430411

- ○ Open daily 10-5 Mar to Oct, 11-3 Nov to Feb.
  Closed Christmas Day and Boxing Day, Jan 1 and 4-8.
- £ Adults £4.00, children £3.00, family £12.00, OAPs £3.50. Group rates (12 or over): 10-15% discount.
- ☞ A5 Britannia Bridge then follow signs along A4080. Bus: from Bangor to Brynsiencyn village (2 miles). Occasional buses call at the zoo during the summer.

Facilities: 🅿 Car and coach parking 🚻 Toilets – parent and baby facilities ♨ Refreshments 🎁 Gift/book shop ⚑ Play area.

Facilities for Disabled: ♿ Wheelchair access.

Restrictions: 🐕 Dogs welcome except in restaurant. Guide dogs welcome anywhere.

Description: The sea zoo is devoted to promoting entirely Welsh marine plants and animals, particularly the sea life around Anglesey. The undercover aquarium depicts natural marine environments in imaginative displays, focusing on duplicating such different areas as harbour walls and under piers. Displays also involve particular subjects such as ecology, fish farming and pollution. There are twenty different seascapes to see; you can experience the feeling of being on the sea bed in a wreck with fish all around or see a demonstration of tides and the power of waves. The Big Fish Forest is an enormous tank with a special curved viewing window providing a spectacular insight into the unique range of sea life to be found in the Menai Straits. A large pool with an innovative use of grid work allows visitors to walk ten centimetres above the water and get stunning views of gurnard with their azure blue fins.

On-Site Activities: Guided tours at extra cost for parties of up to twenty. Beach walk near by. Water games (aquablasta, radio-controlled model boats); adventure playground includes tots' area.

Educational Facilities: Education officer/centre; school visits.

Ellins Tower Seabird Centre

South Stack RSPB Reserve, South Stack, Holyhead, Anglesey, Gwynedd LL65 1TH Tel: (0407) 764973

○ Open daily 11-5 Easter to mid-Sep.

£ Admission free.

☞ A5 to Holyhead and follow signs.
Rail: Holyhead station (2 miles).

Facilities: ▯ Car and coach parking ♦♦ Toilets ⌇ Refreshments.

Facilities for Disabled: ♿ Limited wheelchair access by arrangement.

Restrictions: 🐕 Dogs on leads.

Description: Here is some of the most spectacular sea cliff scenery in Britain, two miles to the west of Holyhead. This RSPB nature reserve covers 780 acres in two parts, in an Area of Outstanding Natural Beauty, and a designated Site of Special Scientific Interest. From the end of April until mid-July, the cliffside ledges are crammed with around 3,500 pairs of breeding guillemots, hundreds of razorbills and dozens of puffins. There are kittiwakes, fulmars, stonechats, ravens and peregrines, and the real speciality here, the chough, a small crow. There are rare plants, too, such as the spotted rock rose, and notably the spathulate fleawort, which flowers in May and June – the best time to visit – and which is found nowhere else in the world. There is a network of footpaths on Holyhead Mountain, and Ellins Tower itself, the visitor centre, is superbly situated on the clifftop overlooking the main seabird colony.

On-Site Activities: Guided walks on Tuesdays and Saturdays May to end August.

Educational Facilities: Education officer/centre; school visits. School groups very welcome; please contact Helen McGarrity, Gallt y Rhedyn, Newborough, Anglesey (0248) 79509.

Pili Palas Butterfly & Bird Palace

Ffordd Penmynydd, Porthaethwy, Menai Bridge, Ynys Môn, Gwynedd LL59 5RP Tel: (0248) 712474

○   Open daily 10-5.30 mid-Mar to end Oct, 11-3 Nov and Dec.

£   Adults £2.95, children £1.75, family £8.00.
    Group rates: adults £2.50, children £1.40.

☞   2 miles from Menai Bridge and Britannia Bridge on B5420 Menai Bridge to Llangefni. Bus: From Bangor to Fourcrosses then walk ¼ mile on B5420.

Facilities: Car and coach parking   Toilets – mother and baby facilities   Refreshments   Gift/book shop   Play area.

Facilities for Disabled: Wheelchair access.

Restrictions: No dogs in play area or inside exhibition.

Description: Predominantly a butterfly farm, this is home to a number of tropical and British butterflies and moths, including several rare breeds. Observe the behaviour and life cycle of the butterflies from egg to caterpillar and from chrysalis to imago. Tarantulas, giant snails and various reptiles, including a garter snake, corn snake and water dragon, can be found in the creepy-crawly cavern and the walk-through aviary holds several captive-bred birds such as the Java sparrow and blue-fronted Amazon parrot. Outside, a nature trail has been set up where twenty different species of British butterfly have so far been recorded. Other attractions include a small mammals corner, peacock pens, duck pond and a dragonfly pond where five species of dragonfly have been identified.

On-Site Activities: Nature trail and dragonfly pond. Talk and thirty-minute video then guided tour of the exhibition for groups of twelve or over.

Educational Facilities: Education officer/centre; school visits.

Snowdonia National Park Visitor Centre & Living Water Aquarium

Betws-y-coed, Gwynedd LL24 0AH Tel: (0690) 710665 (Aquarium 710426)

○ Open daily 10-6 Easter to end Oct, 10-5 in winter. (Aquarium open daily 10-5 Easter to Nov.)

£ Admission free. (Aquarium admission 60p. Unmanned turnstile.)

☞ A5 to Betws-y-coed. Centre is 200 metres off A5 in centre of village opposite Royal Oak Hotel. Bus: Gwynedd network from Llandudno, Porthmagodg, Caernarfon etc. Rail: Betws-y-coed station from Llandudno or Blaenan Ffestiniog.

Facilities: 🅿 Car and coach parking 🚻 Public toilets near by (Aquarium) ☕ Refreshments near by (Aquarium). 👍 Gift/book shop.

Facilities for Disabled: ♿ Wheelchair access.

Restrictions: 🚫 No dogs.

Description: Snowdonia ranges from deep valleys and rugged mountains to rivers, lakes and waterfalls, and from woods and forests to sand dunes and sandy bays. The whole region, which is a stronghold of Welsh language and tradition, has a rich variety of plants and wildlife. The mountains offer superb walking, and, away from Snowdon itself, solitude. Forest and nature trails, for example at the park residential study centre Plas Tan-y-Bwlch, offer easier walks.

The Visitor Centre houses an exhibition on the National Park's landforms, history and present use, and provides an extensive range of activities, information and accomodation services.

The Living Water Aquarium, a private attraction which adjoins the complex, has a fascinating array of freshwater and marine creatures to be found in the area. In all there are seventeen tanks, half fresh and half sea water. Marine displays reflect different habitats and contain lobsters, crawfish, octopuses, grey mullets, sea bass, sea breams and flounders. Freshwater displays cover upland lakes, streams and mountain rivers and contain various plant and fish species including grayling, rainbow trout, ruffe and perch.

On-Site Activities: Guided walks; Welsh sing-along evenings; self-guided trails. Private aquarium of local fish and an RSPB information and exhibition centre as well as a large free national park exhibition.

Special Events: Occasional open days.

Educational Facilities: Education officer/centre; school visits. Lectures available on Snowdonia National Park or Gwydyr Forest (by Forest Enterprise, Llanrwst).

POWYS
Brecon Beacons Mountain Centre

Near Libanus, Brecon, Powys LD3 8ER Tel: (0874) 623366

○ Open daily 9.30-5 Mar to Jun, 9.30-6 Jul to Aug, 9.30-5 Sep to Oct, 9.30-4.30 Nov to Feb.

£ Admission free.

☞ A470 Brecon to Cardiff road; turn west at Libanus 5 miles south-west of Brecon. Bus: Silverline bus, Brecon-Merthyr Fydfu to Libanus then 1¼ miles walk.

Facilities: 🅿 Car and coach parking 🚻 Toilets 🍴 Refreshments 🛍 Gift/book shop ⚠ Play area.

Facilities for Disabled: ♿ Wheelchair access.

Restrictions: 🐕 Dogs on leads.

Description: Impressive views of Pen y Fan and Corn Du, the highest peaks of the Brecons, can be enjoyed from the centre. Whether on a short or long walk there is an abundance of moorland wildlife to look out for: rabbits and hares, voles, shrews, lizards and frogs, toads and newts in wet places. The great variety of insects includes some spectacular dragonflies and damselflies. There is plenty to see for the enthusiastic birdwatcher: nearly 100 different species have been recorded here. Birds to see throughout the year include: heron, sparrowhawk, kestrel, buzzard, lapwing, snipe, curlew, green woodpecker, skylark, mistle thrush, song thrush, goldcrest, meadow pipit, pied wagtail, goldfinch, yellow hammer and reed bunting. Summer visitors include the cuckoo, swift, swallow, house martin, wheatear, whinchat, redstart, willow warbler, chiff chaff, tree pipit, linnet, redshank. Winter visitors are: teal, fieldfare, redwing and brambling. At dawn and dusk keep your eyes peeled for a fox, a stoat or even a polecat.

On-Site Activities: Talks for groups by prior arrangement. Some guided walks, demonstrations etc. at advertised times. Local walks on adjoining common land.

Educational Facilities: Education officer/centre; school visits.

Lake Vyrnwy RSPB Nature Reserve

Bryn Awel, Llandwddyn, Oswestry, Powys SY10 0LZ Tel: (0691) 73278

○ Open daily 10.30-5.30 Apr to end Dec, weekends only Jan to end Mar.

£ Admission free.

☞ Leave A495 about 12 miles south-west of Oswestry at SJ193167, pass through Llanfyllin on B4393 and on to Llanwyddyn.

Facilities: 🅿 Car and coach parking ♦♦ Toilets ☕ Refreshments ♣ Gift/book shop.

Facilities for Disabled: ♿ Wheelchair access; two out of four birdwatching hides have wheelchair access.

Restrictions: 🐕 Dogs on leads.

Description: This is one of the RSPB's largest reserves covering an area of over 16,000 acres of spectacular scenery around the vast Lake Vyrnwy, a reservoir supplying water principally to Liverpool. It has a wide variety of habitats including oak woodlands, conifer forests, meadows, streams and heather moorland, which is itself part of the Berwyn Mountains Site of Special Scientific Interest. There are four birdwatching hides and three nature trails, along with picnic tables. Spring and summer birds include pied flycatchers, redstarts, wood warblers and great crested grebes. Residents include buzzards, ravens, woodpeckers and – of special interest – goosanders, as this is one of the few places in Wales where they breed. They can be watched displaying on the lake early in the year.

On-Site Activities: Guided walks Saturday and Sunday afternoons. Talks available from wardens/teacher naturalist if pre-booked.

Educational Facilities: Education officer/centre; school visits. RSPB teacher naturalist is available to pre-booked groups of children. Visits include a walk round a working farm with a wide variety of livestock and environmental games designed to fulfil specific targets within the national curriculum.

SOUTH GLAMORGAN
Welsh Hawking Centre

Weycock Road, Barry, South Glamorgan CF6 9AA   Tel: (0446) 734687

- ○ Open daily 10.30-5 throughout the year.
- £ Adults £3.00, children £2.00, family £11.00. Group rates: 10% discount.
- ☞ Follow signs for Cardiff airport on A4226 just outside Barry. Bus: to Weycock Cross roundabout, Barry (5 minutes walk).

Facilities: 🅿 Car and coach parking 🚻 Toilets 🍴 Refreshments 🎁 Gift/book shop 🎠 Play area.

Facilities for Disabled: ♿ Wheelchair access.

Restrictions: 🐕 Dogs on leads.

Description: Set in twelve acres of park, this is both a hawking centre and a children's animal park. The hawking centre has a collection of over 200 birds of prey: hawks, eagles, falcons and owls. There are daily flying displays (weather permitting) which include the silent stealth of the beautiful owl, the power and size of the eagle, and the devastating speed and accuracy of the magnificent falcon. The whole display is accompanied by a commentary which not only explains the outstanding features of the bird but also the principles of training, hunting and caring for birds of prey. The children's animal park has pigs, goats and rabbits as well as pony rides and an adventure playground.

On-Site Activities: Flying displays daily (weather permitting).

Educational Facilities: Education officer/centre; school visits.

WEST GLAMORGAN
Penscynor Wildlife Park

Cilfrew, Neath, West Glamorgan SA10 8LF   Tel: (0639) 642189

- ○ Open daily 10-6 Mar to end Sep, 10-4.30 Oct to Feb.
- £ Adults £4.00, children £3.00, OAPs £3.00.
  Group rates (over 20): adults £3.00, children and OAPs £2.00.
- ☞ Follow A465 to Cilfrew, bypassing Neath.
  Bus: No. 158 from Neath.
  Rail: Neath station (2½ miles).

Facilities: Car and coach parking Toilets – disabled and mother and baby facilities Refreshments Gift/book shop Play area.

Facilities for Disabled: Wheelchair access; ramps to café. Toilets. Trails for visually handicapped.

Restrictions: No dogs.

Description: Amongst wild woods, craggy slopes, plunging waterfalls and rushing streams of West Glamorgan's Vale of Neath, is one of Wales' premier collections of exotic wildlife. Within its sixteen acres live over 100 species of exotic bird and twenty-five species of monkey. The centre's growing reputation for breeding rare animals and its commitment to captive breeding of endangered species is gaining national and international recognition. Reptiles, penguins, golden lion tamarins, deer, guanacos, otters and flamingoes are just some of the animals. The zoo centre also provides a range of constantly changing displays and a chance to meet, at very close quarters, a variety of animals including snakes, spiders and stick insects.

On-Site Activities: Guides and talks can be pre-booked.

Educational Facilities: Education officer/centre; school visits. Guides and talks must be pre-booked.

National Tourist Organisations

England Information Line
The England Information Line is a useful service which provides public information on attractions, destinations and places to visit all over England.
Tel: (071) 824 8000

English Tourist Board
Thames Tower, Black's Road, Hammersmith, London W6 9EL
Tel: (081) 846 9000

Isle of Man Department of Tourism, Leisure and Transport
Sea Terminal, Douglas, Isle of Man
Tel: (0624) 674323

Jersey Tourism
Liberation Square, St Helier, Jersey, Channel Islands
Tel: (0534) 78000

Northern Ireland Tourist Board
St Anne's Court, 59 North Street, Belfast BT1 1NB
Tel: (0232) 231221

The Scottish Tourist Board
23 Ravelston Terrace, Edinburgh EH4 3EU
Tel: (031) 332 2433

States of Guernsey Tourist Board
PO Box 23, White Rock, St Peter Port, Guernsey, Channel Islands
Tel: (0481) 726611

The Wales Tourist Board
Brunel House, 2 Fitzalan Road, Cardiff CF2 1UY
Tel: (0222) 499909

England's Regional Tourist Boards

Cumbria Tourist Board
(Covering the county of Cumbria)
Ashleigh, Holly Road, Windermere, Cumbria LA23 2AQ
Tel: (05394) 44444

East Anglia Tourist Board
(Covering the counties of Bedfordshire, Cambridgeshire, Essex, Hertfordshire, Norfolk and Suffolk)
Toppesfield Hall, Hadleigh, Suffolk IP7 5DN
Tel: (0473) 822922

East Midlands Tourist Board
(Covering the counties of Derbyshire, Leicestershire, Lincolnshire, Northamptonshire and Nottinghamshire)
Exchequergate, Lincoln, Lincolnshire LN2 1PZ
Tel: (0522) 531521/3

Heart of England Tourist Board
(Covering the counties of Gloucestershire, Hereford and Worcester, Shropshire, Staffordshire, Warwickshire and West Midlands)
Larkhill Road, Worcester, Worcestershire WR5 2EF
Tel: (0905) 763436

London Tourist Board
(Covering the Greater London area)
26 Grosvenor Gardens, London SW1W 0DU
Tel: (071) 730 3450

Northumbria Tourist Board
(Covering the counties of Cleveland, Durham, Northumberland and Tyne and Wear)
Aykley Heads, Durham DH1 5UX
Tel: (091) 384 6905

North-West Tourist Board
(Covering the counties of Cheshire, Greater Manchester, Lancashire, Merseyside and the High Peak District of Derbyshire)
Swan House, Swan Meadow Road, Wigan Pier,
Wigan WN3 5BB
Tel: (0942) 821222

South-East England Tourist Board
(Covering the counties of East Sussex, Kent, Surrey and West Sussex)
The Old Brew House, Warwick Park, Tunbridge Wells, Kent TN2 5TU
Tel: (0892) 540766

Southern Tourist Board
(Covering the counties of Berkshire, Buckinghamshire, eastern and northern Dorset, Hampshire, Isle of Wight and Oxfordshire)
40 Chamberlayne Road, Eastleigh, Hampshire SO5 5JH
Tel: (0703) 620006

West Country Tourist Board
(Covering the counties of Avon, Cornwall, Devon, parts of Dorset, Somerset, Wiltshire and Isles of Scilly)
60 St David's Hill, Exeter, Devon EX4 4SY
Tel: (0392) 76351

Yorkshire and Humberside Tourist Board
(Covering the counties of Humberside, North Yorkshire, South Yorkshire and West Yorkshire)
312 Tadcaster Road, York YO2 2HF
Tel: (0904) 707961

Wildlife Organisations to Join

Amateur Entomologists' Society
22 Salisbury Road, Feltham, London TW13 5DP
Dedicated to the study and practice of entomology (insects) for juniors and the amateur by publications, correspondence, field meetings and an annual exhibition.

Animal Aid Youth Group
7 Castle Street, Tonbridge, Kent TN9 1BH
Aims to increase public awareness of animal abuse; gives out information, advice and support to young people in their campaign against cruelty to animals.

Bat Conservation Trust
1 Kensington Gore, London SW7 2AR
Exists to protect and encourage a wider understanding of bats. Develops conservation projects. The Young Batworker *newsletter is published once a term.*

British Arachnological Society
Burns Farm, Cornhill, Banffshire AB45 2DL
The society promotes the study of spiders in Britain. Field meetings and talks are regularly attended by schools.

British Deer Society
Beale Centre, Lower Basildon, Reading, Berkshire RG8 9NH
Pools information on deer studies and gives advice on deer management. Regular talks given to children.

British Trust for Conservation Volunteers
36 St Mary's Street, Wallingford, Oxfordshire OX10 0EU
Runs hundreds of working holidays all year round. Open to sixteen to twenty-year-olds. Over 500 affiliated groups. Arranges projects with schools.

Council for the Protection of Rural England
Warwick House, 25 Buckingham Palace Road, London SW1W 0PP
One of the country's leading environmental charities with over 45,000 individual members, active locally, nationally and internationally.

Country Trust
Denham Hill Farmhouse, Quainton, Aylesbury,
Buckinghamshire HP22 4AN
Educational charity; organises expeditions for urban children to the English countryside that may involve getting their hands dirty!

Countryside Education Trust
John Montagu Building, Beaulieu, Brockenhurst,
Hampshire SO42 7ZN
The trust has two field studies centres that enable schoolchildren to explore and enjoy the countryside whilst at the same time gaining a better understanding of the environment. About 7,000 children visit the trust annually.

Donkey Sanctuary
Sidmouth, Devon EX10 0NU
Registered charity concerned with looking after retired donkeys and mules.

Field Studies Council
Central Services, Preston Montford, Montford Bridge,
Shrewsbury, Shropshire SY4 1HW
This runs a number of courses – day, weekend and longer – for individuals and families and has eleven centres throughout England and Wales.

Greenpeace
Canonbury Villas, London N1 2PN
International pressure group set up to act against abuse to the environment. Campaigns against the destruction of wildlife and its environment. Factsheets for children are issued regularly.

The Hawk and Owl Trust
c/o Zoological Society of London, Regent's Park,
London NW1 4RY
Tel: (0223) 892355)
The Hawk and Owl Trust protects wild owls and other birds of prey within their natural habitats. Young members are welcome at a reduced rate and are also encouraged to adopt a nestbox through the special Adopt-a-Box scheme.

International League for the Protection of Horses
Hall Farm, Snetterton, Norfolk NR1E 2LR
Rescues ill-treated horses and restores them to full health. There is an ILPH Junior Club with all manner of activities for fundraising and ILPH involvement.

Mammal Society
Department of Zoology, Bristol University, Woodland Road, Bristol BS8 1UG
Promotes interest in and conservation of British mammals. The youth group organises expeditions that teach children surveying skills.

National Federation of Young Farmers' Clubs
Countryside Department, YFC Centre, Kenilworth, Warwickshire CV8 2LG
YFCs are involved in improving their local countryside whether it be footpath clearance or tree planting and they receive a newsletter three times a year.

National Trust
36 Queen Anne's Gate, London SW1H 9AS
Largest owner of amenity countryside in England, Wales and Northern Ireland with 600,000 acres and 500 miles of coast. Approximately 300 properties open to the public (at a charge) and two million members. It holds more than 1,000 outdoor and indoor public events each year, including special activities for children.

National Trust for Scotland
5 Charlotte Square, Edinburgh EH2 4DU
Scotland's largest voluntary charity organisation with over 230,000 members. Stretching over 100,000 acres, the charity owns over 100 properties ranging from castles to cottages. The education department organises activities for children.

Otter Trust
Earsham, near Bungay, Suffolk
Promotes the conservation of otters and publishes several magazines annually.

Royal Entomological Society
41 Queens Gate, London SW7 5HU
Aims to promote the study and understanding of insects. Supports expeditions and publishes a variety of literature.

Royal Society for the Prevention of Cruelty to Animals (RSPCA)
Causeway, Horsham, Sussex RH12 1HG
Members receive a magazine four times a year and can stand for election to the organisation's council. The junior section for seventeen-year-olds and under, the Animal Action Club, has 70,000 members who receive magazines and can take part in 'care awards' along with Animal Tracks and day trips.

RSNC The Wildlife Trusts Partnership
Witham Park, Waterside South, Lincoln LN5 7JR
A major voluntary organisation with over 250,000 members, this is a partnership of the UK's forty-seven Wildlife Trusts, fifty Urban Wildlife Groups and WATCH, the junior section.

The Scottish Field Studies Association
Kindrogan Field Centre, Enochdhu, Blair Gowry, Perthshire PH10 7PG
Residential courses for all ages at its field centre.

The Wildfowl and Wetlands Trust
Slimbridge, Gloucester GL2 7BT
The Trust has 60,000 members who receive magazines and gain free admission to all centres and also to special members' observatories at Slimbridge and Welney.

World Wide Fund for Nature UK
Panda House, Weyside Park, Godalming, Surrey GU7 1XR
Dedicated to the protection of the living planet, the organisation funds numerous conservation projects at home and abroad. The Go Wild Club specifically caters for the under-eighteens and publishes a variety of literature.

Young Herpetologists Club
c/o Zoological Society of London, Regent's Park,
London NW1 4RY
Anyone with any interest in reptiles and amphibians should join. The YHC organises zoo visits and field trips with an annual camp in Dorset. A newsletter is published four times a year.

Young Ornithologists Club (RSPB)
RSPB Youth Unit, The Lodge, Sandy, Bedfordshire SG12 2DL
This is the youth section of the RSPB. The society owns a number of nature reserves which provide habitats for wild birds. Children are introduced to birds and a number of activities are organised including projects and competitions. Regular publications are issued.

Young People's Trust for the Environment and Nature Conservation
95 Woodbridge Road, Guildford, Surrey GU1 4PY
Concerned with teaching children a practical knowledge of the environment. Courses and lectures organised.

Gazetteer

*References to Wildlife Sites are given in **bold** type, page numbers in italics refer to maps.*

Abbotsbury Swannery,
  Weymouth, Dorset *132-3*, 141
Abernethy Forest Reserve,
  Loch Garten, Nethybridge,
  Inverness-shire *217*, 224
Achnasheen, Ross-shire *217*, 229
Aden Country Park,
  Peterhead, Aberdeenshire *217*, 223
Alfriston, East Sussex *113*, 115
Alnwick, Northumberland *90*, 92
Altrincham, Cheshire *95*, 97
Andover, Hants *132-3*, 148
Anglesey Bird World, Llanfairpwll,
  Anglesey, Gwynedd *241*, 255
Anglesey Sea Zoo, Brynsiencyn,
  Anglesey, Gwynedd *241*, 256
Appleby Castle Conservation Centre,
  Appleby-in-Westmorland,
  Cumbria *11*, 12
Appleby-in-Westmorland, Cumbria *11*, 12
Arundel, West Sussex *103*, 129
Atherstone, Warwicks *56*, 80
Aviemore, Inverness-shire *217*, 225

Bakewell, Derbys *95*, 100
Ballymoney, Co Antrim *209*, 211
Banham Zoo, Norfolk *17*, 37
Barnstaple, Norfolk *158*, 172
Barry, South Glamorgan *241*, 263
Barton-upon-Humber, South Humberside *190*, 193
Basildon Zoo, Basildon, Essex *16*, 29
Bath, Avon *159*, 161
Battersea Park, Children's Zoo,
  London *85*, 86
Beale Bird Park, Reading, Berks *132-3*, 134
Beauly, Inverness-shire *217*, 226
Belfast Zoo, Belfast, Northern Ireland *209*, 210-11
Bempton Cliffs RSPB Nature Reserve, Bridlington, Humberside *190*, 191
Bentley Wildfowl Reserve & Motor Museum, Lewes, East Sussex *113*, 114
Betws-y-Coed, Gwynedd *241*, 259-60
Bewdley, Worcs *56*, 78
Birdland, Bourton-on-the-Water, Glos *56*, 70
Birdworld, Farnham, Surrey *113*, 126
Birmingham Nature Centre,
  Birmingham, West Midlands *56*, 82

Blackpool, Lancs *95*, 102, 106
Blackpool Zoological Gardens,
  Blackpool, Lancs *95*, 102
Blacktoft Sands Nature Reserve,
  Goole, North Humberside *190*, 192
Blair Drummond Safari and Leisure Park, Stirling, Stirlingshire *217*, 219
Blean Bird Park, Canterbury, Kent *113*, 120
Bourton-on-the-Water, Glos *56*, 70, 73
Bovey Tracey, Devon *158*, 172, 178
Brambles Wildlife Park, Herne Bay, Kent *113*, 121
Brean Down Bird Garden, Burnham-on-Sea, Somerset *159*, 182
Brecon Beacons Mountain Centre,
  Brecon, Powys *241*, 260-1
Brentford, Middx *85*, 88
Bridlington, Humberside *190*, 191
Brighton, East Sussex *113*, 118
Brimstone Wildlife Centre, Tregaron, Dyfed *241*, 245-6
Bristol, Avon *159*, 160-1, 162-3
Bristol Zoo Gardens, Bristol, Avon *159*, 160-1
Broadford, Isle of Skye *217*, 228
Broads Authority, The, Norwich, Norfolk *17*, 38
Broads, The, Great Yarmouth, Norfolk *16*, 38
Broadway Tower Country Park,
  Broadway, Worcs *56-7*, 77
Brownsea Island, Poole, Dorset *132-3*, 142
Broxbourne, Herts *16*, 34
Buckfast Butterflies, Buckfastleigh, Devon *158*, 170
Buckfastleigh, Devon *158*, 170
Burford, Oxfords *132-3*, 153
Burnham-on-Sea, Somerset *159*, 182
Butterflies Pleasure Park, Farnsfield, Notts *57*, 65
Butterfly Centre, Eastbourne, East Sussex *113*, 115
Butterfly and Falconry Park, Spalding, Lincs *57*, 59

Cairngorm Reindeer Centre,
  Aviemore, Inverness-shire *217*, 225
California Country Park,
  Wokingham, Berks *132-3*, 136

Camperdown Wildlife Centre, Dundee, Tayside *217*, 238
Canonteign Falls and Country Park, Chudleigh, Devon *159*, 171
Canterbury, Kent *113*, 122, 125
Cardigan, Dyfed *241*, 247
Carnforth, Lancs *95*, 104
Castle Woods Nature Reserve, Llandeilo, Dyfed *241*, 246-7
Causeway Safari Park, Ballymoney, Co Antrim *209*, 211
Chapel-en-le-Frith, Derbys *95*, 99
Chard, Somerset *159*, 183
Charlwood, Surrey *103*, 128
Cheltenham, Glos *56*, 72, 73
Chessington World of Adventures, Chessington, Surrey *103*, 127
Chester Zoo, Upton-by-Chester, Cheshire *95*, 96
Chestnut Centre, Chapel-en-le-Frith, Derbys *95*, 99
Children's Farm, The, Tamworth, Staffs *56*, 79
Chillingham Wild Cattle Park, Alnwick, Northumberland *90*, 92
Cholderton Rare Breeds Farm and Gardens, Salisbury, Wilts *159*, 186
Chudleigh, Devon *159*, 171
Clarences Community Farm, Middlesbrough, Cleveland *90*, 91
Clovelly, Devon *158*, 174
Cluanie Deer Farm Park, Beauly, Inverness-shire *217*, 226
Colchester, Essex *16-17*, 30, 31
Colchester Zoo, Colchester, Essex *17*, 30
Colwyn Bay, Clwyd *241*, 244-5
Cornish Seal Sanctuary, Helston, Cornwall *158*, 164
Cornish Shire Horse Centre, Wadebridge, Cornwall *158*, 165
Corwen, Clwyd *241*, 242
Cotswold Falconry Centre, Moreton-in-Marsh *56*, 71
Cotswold Farm Park, Cheltenham, Glos *56*, 72
Cotswold Wildlife Park, Burford, Oxfords *132-3*, 153
Cotwall End Nature Centre, Dudley, West Midlands *56*, 83
Courage Shire Horse Centre, Maidenhead, Berks 132-3, 137
Cricket St Thomas Wildlife Park, Chard, Somerset *159*, 183
Crossgar, Co Down *209*, 216
Croxteth Hall and Country Park, Liverpool, Merseyside *95*, 109
Cuerden Park Wildlife Centre, Preston, Lancs *95*, 103

Cumbernauld, Strathclyde *217*, 235
Cupar, Fife *217*, 220
Curraghs Wildlife Park, Isle of Man 208

Dartmoor National Park, Bovey Tracey, Devon *158*, 172
Dartmoor Otter Sanctuary, Buckfastleigh, Devon *158*, 170
Dartmouth, Devon *158*, 181
Dedham Rare Breeds Centre, Colchester, Essex *16*, 31
Deep-Sea World, North Queensferry, Fife *217*, 222
Desford, Leics *57*, 58
Dinton Pastures Country Park, Reading, Berks *132-3*, 135
Dorset Heavy Horse Centre,Verwood, Dorset *132-3*, 143
Downpatrick, Co Down *209*, 214-15
Drusillas Park, Alfriston, East Sussex *113*, 115
Dudley, West Midlands *56*, 83, 84
Dudley Zoo and Castle, Dudley, West Midlands *56*, 84
Dulverton, Somerset *158*, 184
Dundee, Tayside *217*, 238
Dunham Massey, Altrincham, Cheshire *95*, 97
Dunstable, Beds *16*, 20

Eastbourne, East Sussex *113*, 115
Easton Farm Park, Woodbridge, Suffolk *17*, 49
Edinburgh Butterfly and Insect World, Lasswade, Midlothian *217*, 230-1
Edinburgh Zoo, Edinburgh *217*, 229-30

Ellins Tower Seabird Centre, Holyhead, Anglesey, Gwynedd *241*, 257
Exmoor Bird Gardens, Barnstaple, Devon *158*, 172
Exmoor National Park, Dulverton, Somerset *158*, 184

Fakenham, Norfolk *17*, 44
Far Ings Nature Reserve, Barton-upon-Humber, South Humberside *190*, 193
Farmer Giles Farmstead, Salisbury, Wilts *159*, 187
Farnham, Surrey *113*, 126
Farnsfield, Notts *57*, 65
Felinwynt Butterflies and Rain Forest Centre, Cardigan, Dyfed *241*, 247

Gazetteer

Flamingo Gardens and Zoological Park, Olney, Bucks *132-3*, 140
Flamingo Land Fun Park, Zoo and Holiday Village, Malton, North Yorks *190*, 195
Folly Farm Waterfowl, Bourton-on-the-Water, Glos *56*, 73
Formby, Merseyside *95*, 110
Formby Red Squirrel Reserve, Formby, Merseyside *95*, 110
Freightliners Farm, Islington, London *85*, 87

Gatwick Zoo, Charlwood, Surrey *103*, 128
Gibraltar Point National Nature Reserve, Skegness, Lincs *57*, 60
Glasgow Zoo Park, Uddingston, Glasgow *217*, 233
Glengoulandie Deer Park, Pitlochry, Perthshire *217*, 239
Godmanchester, Cambs *16*, 25
Goole, North Humberside *190*, 192
Great Yarmouth, Norfolk *16*, *17*, *38*, *39*, *40*, *46*, 48
Guernsey, Channel Islands *205*, 206

Hamerton Wildlife Centre, Huntingdon, Cambs *16*, 24
Hastings, East Sussex *113*, 119
Haverfordwest, Dyfed *241*, 249, 250, 251, 252, 253
Hawk Conservancy, The, Andover, Hants *132-3*, 148
Hayle, Cornwall *158*, 168
Helston, Cornwall *158*, 164
Henfield, West Sussex *103*, 130
Herne Bay, Kent *113*, 121
Heron's Brook Leisure Park and Waterfowl Centre, Narberth, Dyfed *241*, 248
Hexham, Northumberland *90*, 93
High Moorland Visitor Centre, Plymouth, Devon *158*, 173
Highland Wildlife Park, Kingussie, Inverness-shire *217*, 227
Hoddesdon, Herts *16*, 35
Holdenby House Gardens, Holdenby, Northants *57*, 64
Holyhead, Anglesey, Gwynedd *241*, 257
Howletts Zoo Park, Canterbury, Kent *113*, 122
Hunstanton, Norfolk *16*, 43
Huntingdon, Cambs *16*, 24
Hythe, Kent *113*, 123

Ipsden, Oxfords *132-3*, 156
Isle of Man 208

Isle of Skye *217*, 228
Isle of Wight Zoo, Sandown, Isle of Wight *132-3*, 154

Jedburgh, Roxburghshire *217*, 218
Jedforest Deer and Farm Park, Jedburgh, Roxburghshire *217*, 218
Jersey Zoo, Jersey, Channel Islands *205*, 297

Kennet and Avon Canal Visitor Centre, Reading, Berks *132-3*, 138
Kentwell Hall, Sudbury, Suffolk *17*, 50
Kingdom of the Sea, Great Yarmouth, Norfolk *17*, 39
King's Lynn, Norfolk *16*, 47
Kingussie, Inverness-shire *217*, 227
Kinross, Kinross-shire *217*, 240
Knowsley Safari Park, Prescot, Merseyside *95*, 111
Knutsford, Cheshire *95*, 98

Lake District National Park Centre, Windermere, Cumbria *11*, 13
Lake Vyrnwy RSPB Nature Reserve, Oswestry, Powys *241*, 261-2
Lakeland Wildlife Oasis, Milnthorpe, Cumbria *11*, 14
Lasswade, Midlothian *217*, 230-1
Launceston, Cornwall *158*, 169
Le Friquet Butterfly Centre, Guernsey, Channel Islands *205*, 206
Leeds, West Yorks *190*, 202, 203, 204
Leighton Hall, Carnforth, Lancs *95*, 104
Lewes, East Sussex *113*, 114
Lincoln *57*, 63
Lions of Longleat Safari Park, Warminster, Wilts *158*, 188-9
Liverpool, Merseyside *95*, 108, 109
Liverpool Museum Aquarium & Vivarium, Liverpool, Merseyside *95*, 108
Living World, Seaford, East Sussex *113*, 117
Llandeilo, Dyfed *241*, 246-7
Llanfairpwll, Anglesey, Gwynedd *241*, 255
Llyn Brenig Visitor Centre, Corwen, Clwyd *241*, 242
Lodge, The, Sandy, Beds *16*, 18
London *85*, 86, 87, 88, 89
London Butterfly House, Brentford, Middx *85*, 88
London Zoo, Regent's Park, London *85*, 89
Looe, Cornwall *158*, 166

Lotherton Hall Bird Garden, Leeds, West Yorks *190*, 202
Lowestoft, Suffolk *17*, 54
Lullingstone Silk Farm, Sherborne, Dorset *132-3*, 147
Luton, Beds *16*, 23

Maidenhead, Berks *132-3*, 137
Malton, North Yorks *190*, 195
Mansfield, Notts *57*, 67
Margaret Young Home for Animals, Godmanchester, Cambs *16*, 25
Marine Life Centre, St David's, Haverfordwest, Dyfed *241*, 249
Marwell Zoological Park, Winchester, Hants *132-3*, 149
Matlock, Derbys *95*, 101
Mel House Bird Garden, Pickering, North Yorks *190*, 196
Mere Sands Wood Nature Reserve, Rufford, Lancs *95*, 105
Middlesbrough, Cleveland *90*, 91
Milky Way, The, Clovelly, Devon *158*, 174
Milnthorpe, Cumbria *11*, 14
Miniature Pony Centre, Newton Abbot, Devon *158*, 175
Minsmere RSPB Reserve, Saxmundham, Suffolk *17*, 51
Mole Hall Wildlife Park, Saffron Walden, Essex *16*, 27
Monkey Sanctuary, The, Looe, Cornwall *158*, 166
Moors Centre, The, Whitby, North Yorks *190*, 197
Moreton-in-Marsh, Glos *56*, 71
Murlough National Nature Reserve, Newcastle, Co Down *209*, 212

Narberth, Dyfed *241*, 248
National Birds of Prey Centre, Newent, Glos *56*, 74
National Dairy Museum, Reading, Berks *132-3*, 139
National Shire Horse Centre, Plymouth, Devon *158*, 176
National Stud, The, Newmarket, Suffolk *16*, 52
Natural World Centre, Newent, Glos *56*, 75
Natural World, The, Poole, Dorset *132-3*, 144
Neath, West Glamorgan *241*, 264
Nethybridge, Inverness-shire *217*, 224
New Forest Butterfly Farm, Southampton, Hants *132-3*, 150
New Forest Owl Sanctuary, Ringwood, Hants *132-3*, 151
Newcastle, Co Down *209*, 212

Newent, Glos *56*, 74, 75
Newmarket, Suffolk *16*, 52, 61
Northern Ireland Aquarium, Portaferry, Co Down *209*, 213-14
Northern Shire Horse Centre, York *190*, 198
Northumberland National Park, Hexham, Northumberland *90*, 93
Norwich, Norfolk *17*, 38, 42, 45
Nottingham *57*, 68

Oban, Argyll *217*, 234, 237
Oban Rare Breeds Park, Oban, Argyll *217*, 234
Oceanarium, St David's, Haverfordwest, Dyfed *241*, 250
Olney, Bucks *132-3*, 140
Ormskirk, Lancs *95*, 107
Oswestry, Powys *241*, 261-2
Otter Trust, The, Bungay, Suffolk *17*, 53
Owl Centre, The, Ravenglass, Cumbria *11*, 15

Paddock Wood, Kent *113*, 124
Paignton Zoo, Paignton, Devon *159*, 177
Palacerigg Country Park, Cumbernauld *217*, 235
Paradise Park, Hayle, Cornwall *158*, 168
Paradise Wildlife Park, Broxbourne, Herts *16*, 34
Park Farm Tourist Centre, Hunstanton, Norfolk *16*, 43
Parke Rare Breeds Farm, Bovey Tracey, Devon *158*, 178
Paultons Park, Romsey, Hants *132-3*, 152
Peak National Park, Bakewell, Derbys *95*, 100
Peakirk Waterfowl Gardens, Peterborough, Cambs *16*, 26
Pembrokeshire Coast National Park, Haverfordwest, Dyfed *241*, 251
Penscynor Wildlife Park, Neath, West Glamorgan *241*, 264
Pensthorpe Waterfowl Park, Fakenham, Norfolk *17*, 44
Peterborough, Cambs *16*, 26
Peterhead, Aberdeenshire *217*, 223
Pickering, North Yorks *190*, 196
Pili Palas Butterfly and Bird Palace, Ynis Min, Gwynedd *241*, 258
Pitlochry, Perthshire *217*, 239
Plymouth, Devon *158*, 173, 176, 179
Plymouth Aquarium, Plymouth, Devon *158*, 179
Poole, Dorset *132-3*, 142, 144

Gazetteer

Port Lympne Zoo Park, Mansion and Gardens, Hythe, Kent *113*, 123
Portaferry, Co Down *209*, 213-14
Portsmouth, Hants *132-3*, 153
Prescot, Merseyside *95*, 111
Preston, Lancs *95*, 103

Quince Honey Farm, South Molton, Devon *158*, 179-80
Quoile National Nature Reserve, Downpatrick, Co Down *209*, 214-15

Radipole Lake RSPB Nature Reserve, Weymouth, Dorset *132-3*, 145
Ravenglass, Cumbria *11*, 15
Reading, Berks *132-3*, 135, 138, 139
Reg Taylor's Swan Sanctuary, Southwell, Notts *57*, 66
Retford, Notts *57*, 69
Rhyl, Clwyd *241*, 243
Riber Castle Wildlife Park, Matlock, Derbys *95*, 101
Ringwood, Hants *132-3*, 151
Romsey, Hants *132-3*, 152
Royston, Herts *16*, 36
Rufford, Lancs *95*, 105
Rye House Nature Reserve, Hoddesdon, Herts *16*, 35

Saffron Walden, Essex *16*, 27
St Abb's Head National Nature Reserve, St Abb's, Berwickshire *217*, 236
St Andrews, Fife *217*, 221
St David's, Haverfordwest, Dyfed *241*, 249, 250
Salisbury, Wilts *159*, 186, 187
Sandown, Isle of Wight *132-3*, 154
Sandy, Beds *16*, 18
Saxmundham, Suffolk *17*, 51
Scarborough, North Yorks *190*, 199, 200
Scottish Deer Centre, Cupar, Fife *217*, 220
Scunthorpe, South Humberside *190*, 194
Sea Life Centres
    Blackpool, Lancs *95*, 106
    Brighton, East Sussex *113*, 118
    Hastings, East Sussex *113*, 119
    Oban, Argyll *217*, 237
    Portsmouth, Hants *132-3*, 153
    Rhyl, Clwyd *241*, 243
    St Andrews, Fife *217*, 221
    Scarborough, North Yorks *190*, 199
    Weymouth, Dorset *132-3*, 146
Seaford, East Sussex *113*, 117

Seaforde Tropical Butterfly House, Downpatrick, Co Down *209*, 215
Settle, North Yorks *190*, 201
Shaldon Wildlife Trust, Shaldon, Devon *159*, 180
Sherborne, Dorset *132-3*, 147
Sherwood Forest Farm Park, Mansfield, Notts *57*, 67
Skegness, Lincs *57*, 60, 62
Skegness Natureland, Skegness, Lincs *57*, 62
Skokholm Island, Haverfordwest, Dyfed *241*, 252
Skomer Island, Haverfordwest, Dyfed *241*, 253
Skye Serpentarium, Broadford, Isle of Skye *217*, 228
Slimbridge, Glos *56*, 76
Snowdonia National Park Visitor Centre and Living Water Aquarium, Betws-y-Coed, Gwynedd *241*, 259-60
South Molton, Devon *158*, 179-80
Southampton, Hants *132-3*, 150
Southport Zoo and Conservation Trust, Southport, Merseyside *95*, 112
Southwell, Notts *57*, 66
Spalding, Lincs *57*, 59
Spilsby, Lincs *57*, 61
Stagsden Bird Gardens, Stagsden, Beds *16*, 19
Staintondale Shire Horse Farm, Scarborough, North Yorks *190*, 200
Stirling, Stirlingshire *217*, 219
Stonebridge City Farm, Nottingham, Notts *57*, 68
Strumpshaw Fen, Norwich, Norfolk *17*, 45
Sudbury, Suffolk *17*, 50
Suffolk Wildlife and Rare Breeds Park, Lowestoft, Suffolk *17*, 54

Tamar Otter Park and Wild Wood, Launceston, Cornwall *158*, 169
Tamworth, Staffs *56*, 79
Tatton Park, Knutsford, Cheshire *95*, 98
Temple Newsam Home Farm and Estate, Leeds, West Yorks *190*, 203
Thrigby Hall Wildlife Gardens, Great Yarmouth, Norfolk *17*, 46
Titchwell Marsh RSPB Nature Reserve, King's Lynn, Norfolk *16*, 47
Toad Hole Cottage and Wildlife Water Trail, Great Yarmouth, Norfolk *17*, 48

Torridon Countryside Centre, Achnasheen, Ross-shire *217*, 229
Tregaron, Dyfed *241*, 245-6
Tropical Bird Gardens, Bath, Avon *159*, 161
Tropical Bird Gardens, Desford, Leics *57*, 58
Tropical World, Leeds, West Yorks *190*, 204
Tropiquaria, Watchet, Somerset *159*, 185
Twycross Zoo, Atherstone, Warwicks *56*, 80

Uddingston, Glasgow *217*, 233
Ulster Wildlife Centre, Crossgar, Co Down *209*, 216
Upton-by-Chester, Cheshire *95*, 96

Vane Farm Nature Centre and Reserve, Kinross, Kinross-shire *217*, 240
Verwood, Dorset *132-3*, 143

Wadebridge, Cornwall *158*, 165
Warminster, Wilts *158*, 188-9
Washington Waterfowl Park, Washington, Tyne and Wear *90*, 94
Watchet, Somerset *159*, 185
Wellington Country Park, Reading, Berks *132-3*, 139
Wellplace Zoo, Ipsden, Oxfords *132-3*, 156
Welney Wildfowl & Wetlands Centre, Wisbech, Cambs *16*, 27
Welsh Hawking Centre, Barry, South Glamorgan *241*, 263
Welsh Mountain Zoo, Colwyn Bay, Clwyd *241*, 244-5
West Midlands Safari and Leisure Park, Bewdley, Worcs *56*, 78
Westbury, Wilts *159*, 189
Wetlands Waterfowl Reserve, Retford, Notts *57*, 69
Weymouth, Dorset *132-3*, 141, 145, 146
Whipsnade Wild Animal Park, Dunstable, Beds *16*, 20

Whisby Nature Park, Lincoln, Lincs *57*, 63
Whitbread Hop Farm, Paddock Wood, Kent *113*, 124
Whitby, North Yorks *190*, 197
Wicken Fen, Ely, Cambs *16*, 28
Wildfowl & Wetlands Centre, The, Llanelli, Dyfed *241*, 254
Wildfowl & Wetlands Trust, The, Arundel, West Sussex *103*, 129
Wildfowl & Wetlands Trust, The, Martin Mere, Ormskirk, Lancs *95*, 107
Wildfowl & Wetlands Trust, The, Slimbridge, Glos *56*, 76
Willersmill Wildlife Park and Sanctuary, Royston, Herts *16*, 36
Willsbridge Mill, Bristol, Avon *159*, 162-3
Winchester, Hants *132-3*, 149
Windermere, Cumbria *11*, 13
Wingham Bird Park, Canterbury, Kent *113*, 125
Wisbech, Cambs *16*, 27
Woburn Wild Animal Kingdom, Woburn, Beds *16*, 21
Wokingham, Berks *132-3*, 136
Woodbridge, Suffolk *17*, 49
Woodland Leisure Park, Dartmouth, Devon *158*, 181
Woodland Park and Heritage Museum, Westbury, Wilts *159*, 189
Woods Mill Countryside Centre, Henfield, West Sussex *103*, 130
Woodside Farm and Wildfowl Park, Luton, Beds *16*, 23
Worldwide Butterflies and Lullingstone Silk Farm, Sherborne, Dorset *132-3*, 147

Ynis Min, Gwynedd *241*, 258
York *190*, 198
Yorkshire Dales Falconry and Conservation Centre, Settle, North Yorks *190*, 201

Index

*Page references in **bold** type indicate lists of species which are too numerous to index.*

aardvarks 89
adders 51
agoutis 180
alligators 46, 144, 195, 244
alpacas 120
Amazons 58, 168
amphibians 82, 86, 88, 144, 162, 180, 204, 228, 235
anacondas 127
anemones 236, 246
animal sanctuaries and hospitals 25, 36, 58, 62, 66, 68, 69, 99, 104, 146, 151, 164, 165, 166, 174, 196, 204, 207, 219, 226, 237
antelopes 140, 149, 219
ants 21, 82, 117, 230
apes 42, 80, 127, 155
aquaria 14, 39, 62, 89, 106, 107, 108, 109, 112, 144, 160, 163, 167, 179, 195, 213, 221, 222, 237, 243, 249, 250, 256, 259-60
arboreta 71, 223
archaeological trails 238
Areas of Outstanding Natural Beauty 32, 117, 257
asparagus fields 110
Australian bearded dragon 75
aviaries 12, 15, 34, 58, 67, 69, 75, 82, 84, 125, 128, 143, 150, 167, 168, 174, 182, 185, 202, 210, 232, 233, 255, 258
avocets 51, 192

baboons 111
badgers 42, 92, 121, 122, 181, 186, 227, 235, 247, 248
bantams 19, 83, 175, 187, 233
bats 14, 80, 89, 96, 97, 105, 135, 136, 162, 163, 215, 247
bears 22, 30, 167, 244
    Asian black 233
    brown 227, 232, 238
    Himalayan 84
    polar 127, 232
    spectacled 207, 210
beavers 82, 116, 227
bees 59, 75, 82, 130, 162, 180, 181, 215, 230-1
beetles 97, 147, 230
birds 14, 35, 45, 58, 125, 128, 137, 138
    see also seabirds; waterfowl and *individual* genera
birds of prey 19, 20, 62, 64, 74, 100, 101, 104, 124, 126, 128, 135, 151, 174, 176, 195, 196, 201

    see also falcons and falconry
birdwatching 15, 18, 24, 26, 27, 28, 35, 38, 44, 45, 47, 48, 51, 60, 63, 69, 76, 93, 94, 97,105, 107, 129, 141, 145, 191, 192, 193, 214, 224, 235, 242, 254, 260, 261, 262
bison 22, 111, 123, 140, 195, 219, 227
bitterns 35, 38, 51, 193
blackbuck 111
bluebells 246
blueberries 172
boa constrictors 21, 232
bobcats 101
bogland 100, 136, 172, 216, 224
bongos 22
bonobos 80
bramblings 261
buffalo 110, 123, 140, 188
bugs 80
bullfinches 35, 94
buntings 47, 60, 191, 212, 261
bustard, great 20
butterflies 14, 28, 33, **38**, **45**, **48**, 51, 59, 62, **63**,65, 75, 82, 83, 88, 97, **103**, 105, 107,115, 117, 128, **130**, 135, **145**, 147, 150, 170, 176, 188, 192, 204, **206**, 212, **215**, 216, 230, **240**, 245, 247, 258
buzzards 29, 64, 69, 101 104, 121, 174, 176, 184, 226, 242, 261, 262

caddis flies 223
caimans 112
calves *see under* cattle
camellias 141
camels 34, 37, 78, 89, 111, 127, 140, 155, 167, 183, 188, 195, 219, 233
campion, red 197
canals 138
canaries 75
capercaillies 224, 238
capuchins 122, 127, 233
capybaras 112
caracaras 148
carp 36, 50, 58, 65, 138, 245
cassowaries 202, 232
cats, domestic 165
cattle 23, 32, 86, 87, 120, 172, 218
    rare breeds 12, 31, 40, 49, 50, 54, 61, 64, 67, 72, 91, 98, 109, 178, 186, 187, 196, 218, 226, 234
    calves 34, 43, 72, 79, 139, 174, 176, 187, 218, 219, 248
    ankole 78
    belted Galloway 218

British white 50
Chillingham wild white 92
Dexter 175, 218
Highland 77, 239
Jersey 50
Kerry 186
longhorn 49, 72, 198
red poll 49
white park 49, 64
celandines 223
chaffinches 224
chats 42, 212, 236, 257, 261
cheetahs 20, 233
chickens 23, 32, 43, 50, 68, 87, 120, 167, 200
   rare breeds 23, 235
chicks 58, 79, 174, 201
chiff-chaffs 252, 261
children's farms *see under* farms
chimpanzees 30, 33, 37, 80, 84, 96, 112, 127, 210, 219, 232, 244
chinchillas 175
chipmunks 46, 75, 152, 245
choughs 253, 257
clematis 161
clown fish 108
coastline 93, 117, 191, 199, 200, 212, 236, 251, 257
coatis 33, 36, 112, 120, 155, 185
cockatiels 245
cockatoos 70, 125, 128, 140, 161, 168, 173, 182
cod 106
condors 74, 127, 200, 202
conifers 223, 262
conservation work parties 103, 203
coots 63, 103
coral 39, 222
cormorants 38, 48, 97, 145
corn mills 162
cows *see* cattle; farming, dairy
coypus 42
crabs 119, 179, 222
cranes 19, 20, 26, 126, 140, 156, 161
   crowned 19, 22
   European grey 19
   sarus 29, 202
crawfish 259
crocodiles 46, 116, 144, 195
crossbills 224, 242
crustaceans 179, 220
cuckoos 35, 49, 261
curlews 240, 253, 254, 261
cuttlefish 146

dace 138
daffodils 142
dahlias 204

damselflies 28, 63, 97, 103, 105, 138, 260
deer 21, 23, 24, 36, 43, 73, 78, 101, 111, 123, 136, 167, 169, 172, 183, 186, 187, 194, 218, 220, 225, 226, 229, 234, 238, 239, 244, 264
   calves 43
   fallow 82, 84, 92, 97, 98, 121, 139, 218, 220, 226, 238, 247
   hog 220
   muntjac 40, 51, 53, 128, 218, 226, 238
   pampas 22
   Père David's 20, 22, 220, 233
   red 51, 77, 98, 139, 184, 218, 220, 225, 226, 227, 229, 234, 239
   roe 13, 104, 138, 220, 223, 226, 227, 235
   sika 33, 121, 142, 173, 181, 218, 220, 238
   wapiti 78
   water 21
dinosaurs 116, 152, 156
dippers 100, 162, 197, 223
discovery centres 20, 89
dogfish 249
dogs 165
dog's mercury 246
dolphins 62, 164, 195
donkey rides 80, 156, 255
donkeys 22, 23, 77, 80, 86, 124, 154, 156, 164, 175, 219, 255
dormice, fat or edible 82
doves 75
downland 117
dragonflies 18, 28, 48, 63, 97, 103, 105, 130, 135, 138, 150, 192, 258, 260
ducklings 79, 254
ducks 23, 26, 41, 61, 64, 67, 68, 69, 70, 73, 76, 87, 94, 104, 107, 126, 129, 136, 139, 141, 142, 165, 181, 187, 193, 200, 212, 218, 223, 235, 255, 258
   Arctic long-tailed 44
   black East Indian 235
   Carolina 107
   cayuga 235
   harlequin 44, 235
   Indian runner 73, 235, 255
   magpie 235
   mandarin 107, 148
   muscovy 239
   ruddy 107
   silver appleyard 73
   tufted 35, 44, 63, 129, 138
   whistling 26
   white-headed 44
   wood 254, 255
   see also eiders; mallards; pintails; pochards; shelducks; shovelers

eagles 59, 71, 74, 101, 104, 126, 140, 148, 168, 229, 263
   bald 140
   bateleur 201
   golden 71, 201, 232, 237
   steppe 104, 176
echidnas 89
eels
   conger 106, 118, 199, 213, 222, 243
   long fire 108
   moray 153
eiders 26, 44, 94, 107
elands 22, 78
elephants 21, 30, 31, 80, 102, 177, 188, 195, 244
   African 84, 111, 122
   Asian 22, 183
elk, great 110
emus 58, 116, 152
evolution 14
explorers' trails 54

falcons
   lanner 104, 174
   lugger 174
   peregrine 226, 254, 257
falcons and falconry 20, 24, 31, 59, 64, 71, 74,83, 101, 104, 120, 126, 148, 151, 174, 176, 201, 238, 255, 263
   see also birds of prey
farm animals 22, 23, 34, 36, 37, 41, 43, 116, 120
   see also farm animals, rare breeds; farms, children's
farm animals, rare breeds 12, 23, 31-2, 40, 49,50, 54, 61, 64, 67, 69, 72, 79, 91, 98, 101, 109, 125, 176, 186, 187, 196, 218,226, 234, 235, 238, 239 245
farming, dairy 49, 68, 72, 87, 91, 139, 174-5, 187, 248
farming demonstrations 50, 79, 109, 203
   see also horse shoeing; horses, heavy; sheep shearing and dipping
farming history 41, 50, 61, 109, 143, 171, 174, 178, 187, 194, 198, 200
farms
   children's 20, 22, 30, 32, 34, 36, 37, 42, 43, 49, 54, 64, 69, 72, 77, 78, 79, 83, 84, 86, 87, 91, 114, 120, 124, 139, 155-6, 174, 175, 187, 196, 200, 218, 234, 244, 263
   city 68, 87
farmyard trails 43, 72, 203
fenland 27, 28, 38, 45, 48
fieldfares 261
finches 12, 35, 47, 60, 65, 75, 82, 94, 182, 202, 224, 245, 261

fish 14, 18, 36, 39, 65, 82, 89, 106, 119, 144, 146, 155, 179, 195, 205, 222, 236, 237, 245, 249, 256 259, 260
flamingoes 19, 26, 37, 62, 76, 82, 84, 86, 94, 107, 112, 116, 127, 140, 152, 156, 161, 168, 183, 195, 202, 208, 232, 244, 254,264
flat fish 39
fleawort 257
flounders 259
flowers 18, 59, 62, 91, 161, 204, 212, 223, 230, 253
   wild 27, 48, 59, 60, 65, 87, 91, 100, 103, 193, 195, 197, 212, 246, 251
   see also plants; tropical plants
flycatchers 13, 236, 246, 252, 262
forests 67, 80, 224, 227, 242, 259, 262
   see also rainforests; woodland
fossils 154
foxes 36, 82, 89, 92, 99, 101, 226, 229, 247, 261
   Arctic 42, 227, 238
   red 121, 227
   silver 121
frogs 87, 100, 228, 235, 247, 260
   poison-arrow 228
   tree 144, 155
fulmars 117, 191, 236, 257
fungi 240

gadwalls 35, 240
gallinules 19
gannets 126, 191
gardens 13, 18, 46, 67, 68, 75, 87, 96, 104, 109, 116, 128, 140, 141, 152, 155, 160, 161, 166, 170, 172-3, 177, 182, 183, 185, 186, 188, 196, 197, 202, 203, 204, 216
gazelles, Arabian 84
geese 23, 47, 67, 68, 69, 73, 76, 94, 107, 129, 135, 136, 142, 181, 187, 218, 235, 239
   barnacle 26
   bean 45
   Brent 214
   Canada 66, 139, 175
   emperor 94
   greylag 50, 107
   Hawaiian 26, 76, 210, 254
   magpie 107
   Ne Ne 202
   ornamental Chinese 73
   pink-footed 26, 240
   pygmy 44
   red-breasted 70, 107
   Sebastopol 73
   Shetland 235
   snow 12
geology 154, 236

gibbons 24, 46, 80, 160
　　lar 84, 112, 127
giraffes 22, 78, 80, 89, 102, 127, 177, 188, 232
gnus 78
goats 41, 68, 73, 83, 86, 87, 120, 124, 139, 156,176, 187, 234, 239, 255, 263
　　rare breeds 12, 40, 61, 67, 101, 186, 196, 218, 238
　　kids 72, 79, 174, 226, 248
　　Bagot 72, 79, 186
　　golden Guernsey 186, 238
　　Old English 50
　　pygmy 23, 49, 64, 69, 77, 78, 175, 186, 248
godwits 51, 60
goldcrests 13, 103, 261
goldeneyes 13, 105
goldfish 245
goosanders 107, 242, 247, 262
gorillas 80, 102, 122, 123, 127, 188, 207, 210, 232
grassland 18, 26, 38, 63, 97, 103, 145, 216, 252
graylings 260
grebes 45, 240
　　great crested 44, 63, 103, 105, 145, 262
　　little 44, 97, 105
groupers, Queensland 106
grouse 227
guanacos 33, 78, 111, 121, 264
guans 19
guenons 232
guillemots 191, 236, 252, 253, 257
guinea pigs 23, 68, 73, 75, 139, 156, 245, 248
guineafowl 23, 187, 198
gulls 110, 142, 145, 191, 253
gurnard 256

hamsters 25, 46
hares 152, 187, 247, 260
harriers
　　hen 28, 45, 192
　　marsh 28, 38, 45, 48, 51, 192
harvest mice 82, 130
hawks 20, 59, 64, 71, 83, 101, 126, 148, 174, 176, 201, 263
　　see also falcons and falconry
heather 197, 262
heathland 18, 51, 105, 142, 212
hedgehogs 105, 121, 226
hedgerows 72, 91, 135, 138
hens see chickens
herb robert 197
herons 29, 40, 42, 53, 66, 70, 94, 97, 103, 120, 138, 139, 142, 145, 148, 187, 223, 260

herrings 237
hippos 22, 149, 188
hobbys 28, 252
honey production 87, 179-80, 181
　　see also bees
hoopoes 252
hops 124
hornbills 58, 120, 126, 152, 173
horse and cart/wagon rides 40, 61, 137, 143, 150, 165
horse and pony rides 34, 86, 109, 117, 198, 246, 248, 263
horse shoeing 41, 72, 137, 165, 198
horses 23, 25, 49, 50, 52, 67, 98, 198, 245-6
　　heavy 41, 49, 50, 61, 72, 79, 124, 137, 143, 165, 176, 181, 183, 198, 200, 246
　　Przewalski's 20, 78, 123
　　wild 227
hummingbirds 76
huss 249
hyenas, spotted 84

ibis 19, 96, 127, 161, 202, 232
iguanas 21, 155, 215
insects 14, 59, 75, 82, 88, 89, 97, 107, 134, 150,155, 162, 167, 212, 215, 230, 231, 238, 247, 258, 260, 264
iris, yellow 145

jackdaws 117
jaguars 30, 127, 183, 232
jays 13
jellyfish 89, 221
jungle 75, 244

kestrels 69, 83, 104, 121,124, 168, 174, 194, 196, 261
kids *see under* goats
kingfishers 35, 40, 44, 45, 49, 63, 66, 83, 103, 105, 130, 138, 162, 197
kites 64, 101, 148, 202, 226
kittiwakes 191, 236, 253, 257
knots 60

lady's smock 103
lagoons 69, 116, 117, 142, 192
lakes 35, 38, 46, 63, 66, 67, 76, 98, 103, 105, 114, 125, 130, 135, 139, 142, 144, 152, 160, 161, 165, 169, 171, 175, 189, 202, 207, 223, 242, 246, 259, 261-2
lambs *see under* sheep
langurs 122, 232
lapwings 253, 261
lechwe, red 210, 232
lemurs 24, 86, 122, 128, 149, 155, 160, 173, 183, 185, 207, 210, 233
leopard cats 30, 112

leopards 154, 183, 195, 232
   clouded 122
   Persian 155, 160, 244
   snow 30, 37, 46, 123, 127
lichens 100, 225, 246
ling 106
linnets 261
lion fish 153
lions 29, 30, 34, 53, 78, 80, 89, 102, 111, 112,123, 127, 167, 188, 195, 210, 211, 219, 232, 233, 244
liverworts 100
lizards 86, 204, 215, 228, 235, 260
llamas 22, 25, 29, 37, 78, 134, 140, 156, 167, 173, 183, 233
lobsters 179, 222, 249, 259
lorikeets 58, 70, 125
lorys 58, 70, 125
lovage 236
lovebirds 75, 245
lumpsucker fish 146
lungwort 246
lynx 82, 101, 112, 183, 195

macaques 30, 46, 127, 233
macaws 29, 58, 70, 125, 128, 161, 168, 173, 182
magnolias 141
magpies 139
mallards 47, 66, 103, 138, 145, 239, 247
mandrills 112
mangabeys 30
mangroves 228
maras 37, 121, 140, 173
margays 232, 233
marine life *see* aquaria; sea life
marmosets 24, 29, 30, 80, 128, 134, 160, 180, 183, 210
marmots 21
marram grass 110
marshland 45, 47, 51, 53, 60, 107, 117, 130, 135, 142, 193, 254
martins, house 261
meadowland 13, 45, 59, 65, 83, 91, 130, 162, 193, 197, 246, 262
meerkats 24, 37, 86, 116, 155, 180
mice 82, 130, 138
microbes 14
milk-vetch 236
millipedes 14
mills 130, 152, 162
mink 138, 235, 238
mongooses 86, 180
monkeys 22, 34, 36, 37, 53, 80, 111, 116, 123, 128, 149, 155, 156, 166, 173, 177, 180, 183, 188, 208, 232, 233, 244, 264
   Amazon woolly 166
   colobus 30
   Diana 180, 232

grivet 84
patas 84
rhesus 78
saki 180, 232
spider 96, 128
squirrel 29, 128, 180
see also apes; baboons; bonobos; capuchins; chimpanzees ; gibbons; guenons; langurs; lemurs; macaques; mandrills; mangabeys; marmosets; orang-utans; tamarins
moorhens 49, 66, 103, 175, 223
moorlands 100, 172, 173, 184, 197, 224, 235, 242, 251, 260, 262
moose 22
mosses 100
moths 97, 147, 150, 170, 193, 206, 215, 258
mountains 225, 227, 259, 262
mulberries 147
mullet, grey 259
museums and exhibitions 14, 41, 61, 82, 114, 139, 141, 152, 162, 164, 171, 189, 194, 197, 198, 220, 222, 228, 229
see also farming history
mushrooms 100
mynah birds 19

national curriculum 47, 81, 94, 106, 119, 146, 152, 163, 191, 240, 243, 245, 262
National Parks 13, 93, 100, 172, 173, 175, 184,197, 200, 251, 259
nature reserves 27, 63, 69, 78, 105, 114, 130, 136, 142, 162, 192, 193, 200, 212, 214, 219, 224, 236, 239, 240, 246, 251, 261
nature trails 18, 40, 45, 48, 49, 53, 54, 60, 63, 65, 69, 82, 83, 99, 104, 109, 114, 124, 126, 130, 138, 145, 162, 163, 181, 186, 187, 193, 194, 197, 203, 208, 220, 223, 226, 240, 242, 244, 258, 259, 262
newts 87, 235, 260
nightingales 63, 138
nightjars 18, 51
nuthatches 13, 103, 246

oasthouses 124
ocelots 30, 112, 233
octopuses 108, 179, 199, 213, 222, 243, 249, 259
orang-utans 80, 84, 96
orchids 83, 145, 212, 215
oryx 89, 155, 232
ospreys 43, 224, 242
ostriches 126, 244
otters 24, 33, 37, 40, 42, 51, 53, 80, 82,

84, 86, 99, 101, 112, 116, 155, 156, 168, 169, 208, 227, 229, 244, 264
owls 12, 19, 23, 29, 37, 49, 58, 59, 64, 71, 74, 83, 101, 104, 120, 125, 126, 148, 151, 152, 165, 169, 176, 180, 187, 196, 201, 226, 263
   African spotted 196
   barn 15, 19, 28, 42, 44, 75, 99, 116, 121, 151, 165, 174, 196
   boobook 168
   brown fish 15
   eagle 15, 19, 33, 42, 99, 124, 128, 148, 156, 168, 174, 196, 227, 233, 235
   great horned 99
   little 42, 71, 97, 99, 235
   long-eared 235
   scops 15, 148
   snowy 82, 116, 124, 128, 168, 227, 235, 237
   spectacled 148
   tawny 69, 97, 99, 116, 194, 227
oxen 22, 92
oystercatchers 47, 60, 94, 110, 240, 252, 254

pandas 89, 149, 210, 232
panthers 30
parakeets 12, 58, 75, 140, 182
parkland 20, 24, 34, 42, 50, 77, 98, 109, 135, 136, 139, 152, 155, 178, 194, 203, 220, 227, 229, 234, 239, 248
parrots 19, 22, 31, 37, 53, 58, 69, 86, 112, 120, 125, 126, 127, 152, 156, 168, 177, 182, 195, 232, 245, 258
partridges 40, 47, 97, 187
peacocks or peafowl 21, 23, 46, 50, 120, 142, 169, 198, 258
pelicans 62, 126, 140, 152, 173, 232
penguins 30, 31, 37, 62, 70, 80, 102, 112, 116, 126, 127, 156, 161, 164, 167, 168, 173, 210, 232, 244, 264
perch 260
pet animals 22, 25, 29, 37, 42, 43, 49, 59, 62, 67, 68, 72, 73, 83, 91, 111, 112, 125, 139, 143, 152, 161, 165, 176, 178, 186, 188, 200, 219, 234, 245, 246, 248
petrels, storm 252
pheasants 12, 19, 23, 40, 41, 64, 97, 120, 126, 156, 187, 196, 232
   bar-tailed 19
   cheer 19, 42
   golden 53, 64, 169
   Lady Amhurst 64
   ornamental 161, 248
   silver 64
   Szechwan white-eared 46
   yellow golden 64
pigeons, pink 168, 207
piglets 64, 152, 219

pigs 12, 23, 34, 43, 61, 67, 68, 77, 83, 87, 109, 120, 139, 175, 176, 178, 187, 196, 234, 255, 263
   rare breeds 31, 40, 61, 68, 79, 98, 109, 194, 218
   Berkshire 218
   black 40, 218
   Gloucester Old Spot 32, 40, 49, 64, 72, 218
   lops 40
   middle white 218
   Oxford sandy 40
   pot-bellied 29, 33, 41, 64, 78, 134, 245, 248
   saddleback 40, 79, 218
   Tamworth 40, 50, 218
pike 138
pine martens 36, 226, 227, 229, 237, 244
pintails 105, 254
pipefish 146, 249
pipits 18, 191, 240, 253, 261
pirate trails 39
plants 62, 107, 145, 171, 254, 257
   see also flowers; tropical plants
plovers 47
pochards 13, 44, 129, 247
polecats 244, 261
pollack 106
pond skaters 223
pond-dipping 27, 91, 93, 130, 162, 192, 203
ponds 12, 18, 59, 63, 83, 101, 103, 114, 116, 150, 152, 162, 165, 166, 170, 180, 213, 216, 229, 230, 258
ponies 41, 50, 61, 78, 143, 156, 165, 171, 172, 175, 178, 184, 187, 200, 219, 246
pony rides, horse and 34, 86, 109, 117, 188, 246, 248, 263
porcupines 29, 36, 46, 233
poultry 12, 41, 61, 67, 98, 109
   rare and old breeds 19, 23, 128, 178, 196, 218, 234, 235
practical hands-on exhibits 14, 20-1, 81, 116, 119, 162
praying mantids 150, 230
primroses 246
puffins 191, 236, 252, 253, 257
pumas 29, 154, 167, 195
pythons 75, 144, 167, 195, 215, 233

quail 170, 247

rabbits 23, 36, 68, 73, 75, 77, 82, 101, 104, 120, 138, 139, 156, 165, 167, 175, 176, 178, 186, 187, 218, 219, 234, 245, 247, 248, 260, 263
racoons 29, 36, 101, 121
rainforest 21, 37, 65, 116, 146, 166, 185, 231, 245, 247

rare breeds of farm animals
  see under cattle; chickens; farm
  animals; goats; pigs; poultry; sheep
rats, black 89
ravens 253, 257, 262
rays 39, 118, 153, 213, 221, 243, 249, 250
razorbills 191, 236, 252, 253, 257
redpolls 240, 242
redshanks 51, 240, 254, 261
redstarts 246, 261, 262
redwings 261
reed mace 145
reedbeds 27, 28, 35, 45, 47, 48, 51, 130, 141, 145, 192, 193
reedlings, bearded 28
reef fish 106
reindeer 42, 177, 220, 225, 227, 238
reptiles 37, 75, 80, 82, 84, 88, 89, 111, 112, 117, 127, 134, 144, 160, 167, 177, 180, 185, 195, 207, 215, 228, 232, 233, 244,258, 264
reservoirs 242, 262
rheas 86, 112, 126
rhinos 149, 188, 232
  black 89, 96
  Indian 20
  Sumatran 123
  white 20, 22, 30, 111, 155, 233
rhododendrons 141
rivers and streams 38, 48, 49, 70, 83, 103, 135, 162, 164, 169, 176, 197, 214, 242, 246, 259, 262
rock roses 257
roses 204, 212
RSPB Nature Reserves 18, 35, 47, 50, 51, 145, 191, 224, 257, 261-2
rudd 138
ruffe 260

safari parks 78, 111, 115, 188, 211, 219
safari rides 43, 123, 127, 188, 219
salamanders 46, 155
sanderlings 110
sandpipers, green 35
saxifrage, golden 197
scallops 213
scorpions 59, 117, 150, 215, 230
sea anemones 14
sea bass 259
sea breams 259
sea buckthorn 212
sea lavender 60
sea life 39, 60, 62, 106, 108, 118, 119, 146, 153, 199, 221, 222, 237, 243, 249, 250, 256
sea mouse 119
sea scorpions 249
sea urchins 211
seabirds 60, 62, 117, 191, 236, 251, 252, 257
sealions 20, 31, 78, 80, 102, 111, 162, 183, 188,195, 208, 210, 219, 232, 244
seals 31, 37, 62, 80, 162, 212, 213, 221, 237, 251, 253
seashore 39, 60, 117, 126, 179, 199, 212, 229, 251
secretary birds 74, 126
sedge beds 48, 145
serval cats 33, 112, 156, 183
shags 126, 191, 236
sharks 39, 106, 109, 118, 144, 153, 199, 222, 243, 249, 250
shearwaters 236, 252, 253
sheep 23, 43, 61, 67, 68, 69, 83, 87, 124, 139, 156, 172, 181, 255
  rare breeds 12, 32, 40, 72, 101, 109, 186, 196, 218, 234, 235, 238, 239, 245
  lambs 32, 34, 43, 64, 68, 72, 152, 174, 176, 187, 218, 219, 226, 234, 248
  Barbary 78, 84
  Castlemilk Mourit 186
  Cotswold 77
  Gotland 218
  grey-faced Dartmoor 218
  Hebridean 186, 218, 235
  Jacob 235
  Lewis blackface 235
  Lincoln longwool 50
  Manx Loghtan 218
  Mouflon 218
  Norfolk horn 50, 186
  North Ronaldsay 235
  Portland 186
  Shetland 218, 235
  Soay 64, 72, 79, 82, 218, 235, 238
  South Down 218
  wild 140
sheep shearing and dipping 32, 43, 68, 72, 79, 82, 87, 226, 235
sheepdogs and shepherding 91, 141
shelducks 26, 47, 240
shovelers 35, 105, 107, 247
shrews 138, 260
shrimps 199
silk farming 128
Sites of Special Scientific Interest 105, 136, 224, 252, 254, 257, 262
skinks 232
skuas 236, 252
skylarks 47, 261
sloths 24
smew 129
snails, giant 89, 96, 258
snake-handling 31, 75, 154, 185, 228
snakes 65, 75, 78, 86, 101, 112, 127, 144, 155, 167, 185, 195, 204, 215, 228, 233,

258, 264
snipe 28, 261
softbills 58, 125, 128
sparrowhawks 63, 83, 94, 97, 194, 261
sparrows 191, 258
spiders 14, 21, 59, 65, 75, 89, 101, 108, 112, 117, 144, 150, 154, 180, 185, 204, 215, 230, 245, 258, 264
spoonbills 252
squirrels 18, 36, 92, 121, 135, 136, 139, 162, 186
   grey 184, 197, 247
   red 13, 105, 110, 142, 227, 244
starfish 14, 119, 199, 213, 243
starlings, tropical 202
steer, Highland 227
stick insects 59, 89, 117, 147, 150, 230, 247, 264
sticklebacks 146
stingrays 199
stints, little 51
stoats 99, 187, 235, 261
stonechats 51, 212, 257
storks 127, 140
streams *see* rivers and streams
suckerfish 249
swallows 47, 145, 261
swamps 46
swans 23, 25, 26, 27, 66, 67, 69, 76, 94, 107, 129, 135, 136, 139, 141, 212
   Bewick's 26, 27, 76, 107, 129
   black 121, 255
   black-necked 26
   mute 141, 145
   trumpeter 26, 94, 148
   whooper 27, 107, 214, 240
swifts 252, 261

tamarins 30, 80, 89, 160, 180, 185, 207, 210, 233, 264
tanagers 155
tapirs 112
tarantulas 21, 59, 65, 75, 89, 101, 108, 144, 150, 180, 215, 230, 245, 258
teals 105, 107, 214, 247, 254, 261
tench 138
terns 35, 47, 51, 60, 142, 214
terrapins 170, 215, 245
theme parks 127
thrushes, song and mistle 261
tigers 22, 53, 78, 102, 111, 127, 195, 211, 232, 233
   Bengal 154, 155
   Indian 122, 123
   Siberian 30, 49, 122, 123, 219
   Sumatran 46, 49, 80, 89, 160
   white 188
tits 194
   bearded 51, 145, 192, 193

blue 197
crested 224
long-tailed 18, 49, 103
toads 110, 235, 247, 260
   natterjack 18, 100
toadstools 100
tortoises 25, 155, 185, 228, 233
toucans 58, 86, 140, 156, 168
tractor and trailer rides 43, 67, 218
tree creepers 103, 246
tropical birds 58, 62, 65, 70, 76, 128, 146, 152, 155, 160, 161, 172-3, 182, 183, 185, 244, 264
tropical fish 39, 65, 106, 134, 185, 204
tropical plants 59, 65, 115, 128, 141, 147, 155, 160, 183, 185, 204, 230
trout, rainbow 260
turacos 202
turkeys 23, 32, 41, 43
turtles, loggerhead 213
twites 47

violas 204
voles 51, 138, 253, 260
vultures 74, 101, 126, 140, 148, 168, 201

waders 45, 47, 51, 109, 142, 212, 214
wagtails 145, 261
wallabies 21, 24, 29, 42, 53, 69, 78, 101, 111, 112, 116, 120, 121, 140, 156, 167, 169, 183, 219
walruses 62
warblers 194, 236
   cetti's 145
   garden 13
   reed 35, 145, 193
   sedge 253
   willow 240, 252, 261
   wood 13, 262
wasps 75
water dragon 258
water gardens 67
water lilies 48
water mills 130, 152
water rails 51, 193
water trails 48
waterfalls 65, 101, 116, 169, 170, 171, 230, 245, 259
waterfowl 12, 19, 23, 26, 36, 44, 48, 53, 58, 66, 69, 70, 73, 75, 76, 94, 105, 107, 114, 125, 139, 140, 161, 171, 175, 181, 183, 202, 219, 246, 248, 254
weasels 235
weaver birds 65
wetlands 27, 38, 69, 76, 107, 117, 129, 145, 208, 214, 216, 246, 254
whales 14, 62, 164
wheatears 261
whinchats 261

white-eyes 155
whitethroats 45, 193, 212, 252
wigeons 214, 247, 254
wildcats 24, 82, 121, 237
wildebeest 22, 53, 111
wolf fish 236
wolves 21, 36, 123, 195
   Canadian timber 30, 78, 238
   maned 37
wood mice 138
woodcock 28
woodland trails 44, 49, 114, 126, 224, 244, 245
woodland walks 34, 37, 54, 58, 77, 97, 136, 138, 150, 202, 233, 248
woodlands 12, 13, 18, 26, 28, 48, 49, 51, 58, 63, 65, 83, 99, 100, 103, 109, 110, 114, 121, 126, 130, 139, 141, 142, 150, 161, 162, 180, 181, 189, 194, 105, 197, 203, 212, 216, 223, 227, 235, 238, 242, 244, 246, 248, 251, 262, 264

woodpeckers 13, 18, 35, 49, 136, 139, 194, 262
   great spotted 83, 94, 97, 103, 121, 130, 246
   green 83, 97, 121, 246, 261
   lesser spotted 97, 246
working models 14, 116
wrasse, Australian cleaner 106

yaks 78
yellow surgeon fish 108
yellowhammers 261

zebras 22, 34, 37, 53, 78, 111, 149, 155, 183, 219, 232
zeedonks 30
zoos 14, 20-1, 29, 30-1, 37, 78, 80-1, 82, 84, 86, 89, 96, 102, 112, 115-16, 117, 120, 122, 123, 127, 128, 140, 149, 154, 156, 160-1, 167, 177, 195, 207, 210-11, 231-2, 233, 244-5, 256, 264

Other titles from BBC Children's Books:

Wildside on Grasslands	*Susan MacMillan*
Wildside on Oceans	*Nicola Davies*
Wildside on Polar Regions	*Paul Appleby*
Wildside on Rainforests	*Paul Appleby*
Wildside on Rivers, Lakes and Wetlands	*Susan MacMillan*
Wildside on Woodlands	*Tess Lemon*
Blue Peter Action Book	*Lewis Bronze and Peter Brown*
Blue Peter Green Book	*Lewis Bronze, Nick Heathcote and Peter Brown*
Brunei Rainforest Adventure	*Peter Brown*